Data-Driven Modeling: Using MATLAB® in Water Resources and Environmental Engineering

Water Science and Technology Library

VOLUME 67

For further volumes:
http://www.springer.com/series/6689

Shahab Araghinejad

Data-Driven Modeling: Using MATLAB® in Water Resources and Environmental Engineering

Springer

Shahab Araghinejad
College of Agriculture and Natural Resources
Irrigation and Reclamation Engineering
University of Tehran
Tehran, Iran

Additional material to this book can be downloaded from http://extras.springer.com

ISBN 978-94-024-0589-7 ISBN 978-94-007-7506-0 (eBook)
DOI 10.1007/978-94-007-7506-0
Springer Dordrecht Heidelberg New York London

Softcover reprint of the hardcover 1st edition 2014

Printed on acid-free paper

Springer is part of Springer Science+Business Media (www.springer.com)

Dedicated to my wife Neda
and my daughter Roz

Preface

The purpose of writing this book has been to give a systematic account of major concepts and methodologies of data-driven models and to present a unified framework that makes the subject more accessible and applicable to researchers and practitioners. The book is structured to integrate important theories and applications on data-driven models and to use them in a wide range of problems in the field of water resources and environmental engineering. The presented models are useful for various applications, namely, hydrological forecasting, flood analysis, water quality monitoring, quantitative and qualitative modeling of water resources, regionalizing climatic data, and general function approximation. This book addresses the issue of data-driven modeling in two contexts. Theoretical background of the models and techniques is presented and discussed in a comparative manner, briefly. Also the source files of relative programs demonstrating how to use the explained models are presented with practical advice on how to advance them. The programs have been developed within the unified platform of MATLAB. The proposed models are applied in various illustrative examples as well as several workshops. The focus of the book remains a straightforward presentation of explained models by discussing in detail the necessary components and briefly touching on the more advanced components.

The book is served as a practical guide to the main audience of graduate students and researchers in water resources engineering, environmental engineering, agricultural engineering, and natural resources engineering. This book may also be adapted for use as a senior undergraduate and graduate textbook by selective choice of topics. Alternatively, it may also be used as a resource for practicing engineers, consulting engineers, and others involved in water resources and environmental engineering.

The book contains eight chapters; except first and last, each was developed in two parts of theory and practice to achieve the aim of the book.

Chapter 1 lays the foundation for the entire book with a brief review on different types of models that could be used for modeling water resources and environmental problems as well as the process of model selection for a specific problem. Furthermore, the general approach of using data-driven models is reviewed in this chapter.

Chapter 2 presents discrete and continues probability distribution functions. Since one of the most applicable fields of distribution functions is frequency analysis, dealing with this issue is also presented in this chapter. The hypothetical tests on the average and variance of one and two populations are reviewed in the chapter. Furthermore, two famous tests of *chi-square* and *Kolmogorov–Smirnov* are presented to decide on the best distribution function for a specific random variable. Each of the above calculations is supported by related commands provided in MATLAB.

Chapter 3 presents models for point and interval estimation of dependent variables using different regression methods. Multiple linear regression model, conventional nonlinear regression models, logistic regression model, and *K*-nearest neighbor nonparametric model are presented in different sections of this chapter. Each model is supported by related commands and programs provided in MATLAB.

Chapter 4 focuses on methods of time series analysis and modeling. Preprocessing of time series before being used through modeling containing assessment of different components is discussed in this chapter. Autoregressive (AR), autoregressive moving average (ARMA), autoregressive integrated moving average (ARIMA), and autoregressive moving average with eXogenous data (ARMAX) models are presented and discussed in this chapter. A review on the multivariate analysis of time series modeling is added. Two major applications of time series modeling, namely, forecasting and synthetic data generation, are presented, supported by the related syntaxes and programs in MATLAB.

Chapter 5 deals with the artificial neural networks (ANNs). It presents basic definition on the "Components of an ANN," "Training Algorithm," and "Mapping by ANNs." The chapter also deals with the introduction of famous ANN models, which are widely used in different fields. Theoretical background, network architecture, their training and simulation methods, as well as the codes necessary for applying the networks are presented in this section. After presenting each network sample, applications are discussed in different illustrative examples. In this chapter, static and dynamic networks and also statistical networks are those which are described and modeled in MATLAB.

Chapter 6 presents the concept of support vector machine (SVM) to analyze data and recognize patterns, required for classification and regression analysis. Two applications of SVMs including classification and regression are discussed in this chapter, and examples on using SVM for both mentioned purposes are presented. Examples and models of SVM are presented in MATLAB.

Chapter 7 is concerned with the fuzzy logic. Basic information in fuzzy logic, fuzzy clustering, fuzzy inference systems, and fuzzy regression are the main subjects presented in this chapter. Obviously, the related MATLAB commands are presented to support the models reviewed in this chapter.

Chapter 8 begins with a summary on the characteristics of the models presented in the previous chapters. The models are compared based on different criteria to give the readers ideas on how to take advantages of the models' strengths and avoid their weaknesses through the hybrid models and multi-model data fusion approach. The chapter continues with the examples of hybrid models and general techniques of multi-model data fusion.

The book also contains an appendix that helps readers to use MATLAB.

I would like to express my gratitude to my colleagues and students who helped me to complete and enhance this book. I very much thank Neda Parvini for providing artworks and graphics of the book and also her invaluable support and encouragement during this project. I would like to thank Shahrzad Farzinpak who has contributed effectively by reviewing the book. I greatly appreciate the patience and concern of Petra van Steenbergen, senior publishing editor, and Hermine Vloemans at Springer. The comments on different chapters of the book received from M. Moghaddas, H. Rousta, E. Meidany, and M. Farhangi are highly acknowledged.

Acknowledgement Constructive comments received from Mr. E. Bozorgzadeh are highly appreciated.

Tehran, Iran
2013

Shahab Araghinejad

Contents

Chapter 1
Introduction

Abstract Problems involving the process of water resources and environmental management such as simulation of natural events, warning of natural disasters, and impact analysis of development scenarios are of significant importance in case of the changing environment. Considering the complexity of natural phenomena as well as our limited knowledge of mathematical modeling, this might be a challenging problem. Recently, development of data-driven models has improved the application of specific tools to be used through the complex process of real-world modeling. Soft computing and statistical models are two common groups of data-driven models that could be employed to solve water resources and environmental problems. Data-driven models are among mathematical models, which use experimental data to analyze real-world phenomena. In contrast to physical models, they do not need a specific laboratory setup so are significantly cheaper. Also, in contrast to the analytical models, data-driven models can be used for the problems where we do not have enough knowledge about the intrinsic complexity of the phenomena. This chapter presents a brief review of different types of models that could be used for modeling water resources and environmental problems, reviews the process of model selection for a specific problem, and investigates the general approach of using data-driven models. The advanced stage of developing a model is discussed in the last section.

Keywords Data-driven models • Model selection • Type of models • Type of data • Decision support systems

1.1 Introduction

Modeling a system is one of the most significant challenges in the field of water resources and environmental engineering. That rise as a result of either physical complexity of a natural phenomenon or the time-consuming process of analyzing different components of a system. Data-driven models have been found as very

S. Araghinejad, *Data-Driven Modeling: Using MATLAB® in Water Resources and Environmental Engineering*, Water Science and Technology Library 67,
DOI 10.1007/978-94-007-7506-0_1, © Springer Science+Business Media Dordrecht 2014

powerful tools to help overcoming those challenges by presenting opportunities to build basic models from the observed patterns as well as accelerating the response of decision-makers in facing with the real-world problems. Since they are able to map causal factors and consequent outcomes of an event without the need for a deep understanding of the physical process surrounding the occurrence of an event, these models have become popular among water resources and environmental engineers. Also, as recent progresses in soft computing have enriched the collection of data-driven techniques by presenting new models as well as enhancing the classic ones, continuity of such popularity is expected.

Data-driven models, as it is understood by its name, refer to a wide range of models that simulate a system by the data experienced in the real life of that system. They include different categories generally divided into statistical and soft computing (also known as artificial-intelligent) models. Data-driven models are often inexpensive, accurate, precise, and more importantly flexible, which make them able to handle a wide range of real-world systems with different degrees of complexity based on our level of knowledge and understanding about a system. As far as the statistical type of them is concerned, these models could be considered among the very primary models in the life of modern engineering. However, they could be categorized as brand new models with regard to soft computing type. Data-driven modeling could be defined as a solution defined by the paradigm of "engineering thinking and judgment" to the world of modeling to deal with the problems, which are considered too complex by our knowledge of mathematical equations. Data-driven models have been brought up to their present form by the ideas and applications from different fields of engineering.

As complexity of a system increases, efficiency of offered models by data-driven methods rises in modeling the system. For systems with little complexity, analytical models based on mathematical equations provide precise descriptions, but for the ones with significant complexity, data-driven models are more useful to define the patterns within the behavior of the system.

Data-driven models can be applied in a wide range of problems including simulation of natural phenomena, synthetic data generation, forecasting and warning of extreme events, developing decision-making rules, and many others. Generally, the purpose of data-driven modeling includes but is not limited to:

- Data classification and clustering
- Function approximation
- Forecasting
- Data generation
- General simulation

As far as problems in the field of water resources and environmental engineering are concerned, application of data-driven models may include:

- Water quality simulation and prediction
- Extreme value prediction with emphasis on floods and droughts

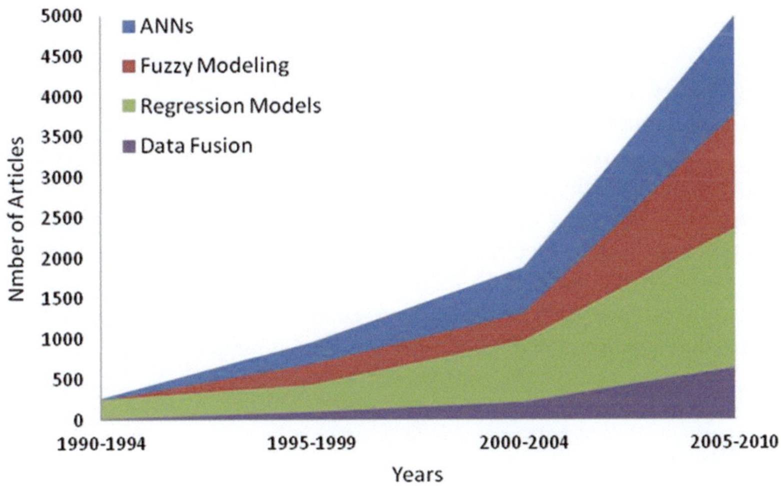

Fig. 1.1 Number of articles published in the selected data-driven techniques (Google Scholar)

- Modeling water balance concerning different components of a hydrological system
- Extending the length of hydroclimatological data from the historical ones
- Estimating censored data

The inexpensive process of developing data-driven models makes them a good choice as either the main tool for modeling a system or an alternative for the baseline model to be compared with the results obtained by analytical and physical models to validate them or to provide useful data and information to enhance them.

There has been an increasing interest on the data-driven modeling in the field of water resources and environmental engineering during the recent decade. Figure 1.1 demonstrates the trend of the number of articles published in the selected data-driven techniques including artificial neural networks (ANNs), fuzzy modeling, regression models, and data fusion in the field of water resources and environmental engineering in the period of 1990–2010. However, the presented statistic might not represent the actual number of researches in those fields; their relative changes demonstrate the fact of an increasing demand on the application of those models. The figure shows a considerable ascending trend in all classic and modern techniques.

This introductory chapter starts with a section discussing different types of models and their complexity. The chapter continues by presenting criteria to select a model. Despite the difference in the type of data-driven models, they all follow a general approach of modeling, which is presented in another section of this chapter. The chapter ends by a discussion on the next level of developing computing tools that could be imagined in the field of modeling and simulation.

1.2 Types of Models

The term "model" refers to a wide variety of programs, softwares, and tools used to represent a real-world system in order to predict its behavior and response to the changing factors approximately. In spite of the wide number, models are divided into two main categories: physical and mathematical models. Data-driven models are classified among the latter category. Figure 1.2 shows a proposed framework for different types of models, specifically those which could be used in the field of water resources and environmental field. As the figure depicts, a mathematical model is in turn divided into three types of data-driven, conceptual, and analytical models. The next expressions present a brief description of shown models in Fig. 1.2.

1.2.1 Physical Model

A physical model is a smaller or larger physical copy of a system. The geometry of the model and the object it represents are often similar in the sense that one is a rescaling of the other.

1.2.2 Mathematical Model

A mathematical model is based on mathematical logic and equations, which benefits from mathematical knowledge to simulate a system in an explicit or implicit manner. It is presented in the forms of analytical, conceptual, and data-driven models.

1.2.3 Analytical Model

An analytical model is a model that represents a system by mathematical equations explicitly. These models are applied in cases that are not too complex comparing to our knowledge of mathematics. Models developed for porous media and groundwater environment are examples of this kind of model.

1.2.4 Data-Driven Model

Data-driven models also known as *experimental models* refer to a kind of models which benefit input and output data of a system to find out specific patterns to be generalized for a broader range of data. Statistical models and artificial-intelligent models are two famous types classified in this category.

Fig. 1.2 A general classification of different types of models

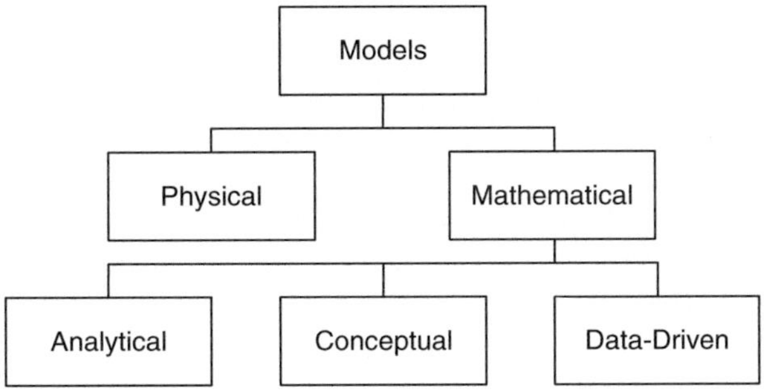

Fig. 1.3 Status of three types of mathematical models based on two characteristics of complexity and mathematical knowledge

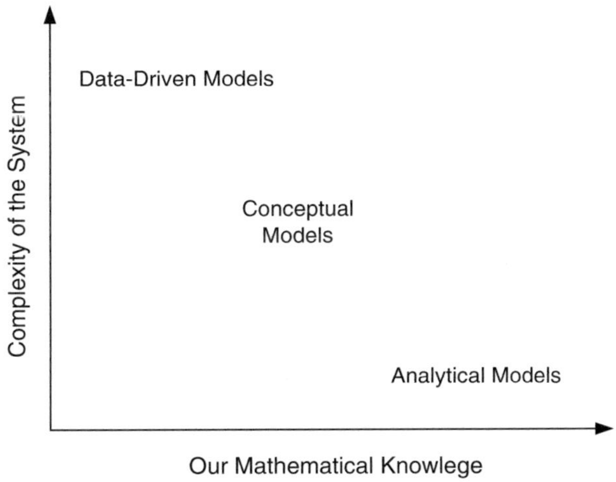

1.2.5 Conceptual Model

A conceptual model in water resources and environmental engineering is a model that benefits from the physical definition of a system partially in cases that we do not have adequate knowledge of mathematical equation to represent the system by an analytical model. A conceptual model uses a combination of both analytical and data-driven model approaches in a way to employ the benefits of both to simulate a system.

Based on the presented definition, different types of mathematical models could be classified by the complexity of a system as well as our mathematical knowledge, as shown in Fig. 1.3.

Data-driven models as the focus of this book are divided into two general forms of statistical models and soft computing models defined as follows.

1.2.5.1 Statistical Model

The definition of a statistical model is tied to the definition of a stochastic process. A stochastic process includes both random and deterministic variables. Deterministic variables are dealt with mathematical models. Meanwhile, a random variable is represented by the theory of probability and a probabilistic model. A probabilistic model is a probability distribution function proposed as generating data. To simulate a stochastic process, statistical models, which benefit from both mathematics and probability theory, are used to model stochastic variables. These models could be parametric or nonparametric. The former has variable parameters, such as the mean and variance in a normal distribution. The latter has a distribution function without parameters, such as nearest neighborhood modeling, and only loosely confined by assumptions.

1.2.5.2 Soft Computing Model

Fuzzy logic, neuro-computing, and genetic algorithms may be viewed as the principal constituents of what might be called soft computing. Unlike the traditional hard computing, soft computing accommodates the imprecision of the real world by the rules obtained from the biological concepts.

1.3 Spatiotemporal Complexity of a Model

Regardless to the type of a model, it could be classified by its complexity in a spatial and temporal manner. The significance of dynamic change of natural phenomena related to water resources and global environment along with the need to asses them in a regional manner make these characteristics of great importance for a modeler. Even though, a fully developed dynamic and distributed model is considered as a very powerful tool in the field of water resources and environmental engineering, the limitation of data, the level of our expectation from the model, as well as the complicated and time-consuming process of modeling might prevent application of such a model in every problem we deal with. The following expressions define different types of models from the spatiotemporal complexity aspect.

1.3.1 Spatial Complexity

From the spatial complexity aspect, a model is categorized as lumped, distributed, and semi-distributed model. Variables in water resources and environmental engineering are usually defined within a specific spatial boundary such as watersheds, basins, provinces, and rivers. A lumped model is used in cases that the spatial variation of this variable within that boundary is not of interest of the modeler.

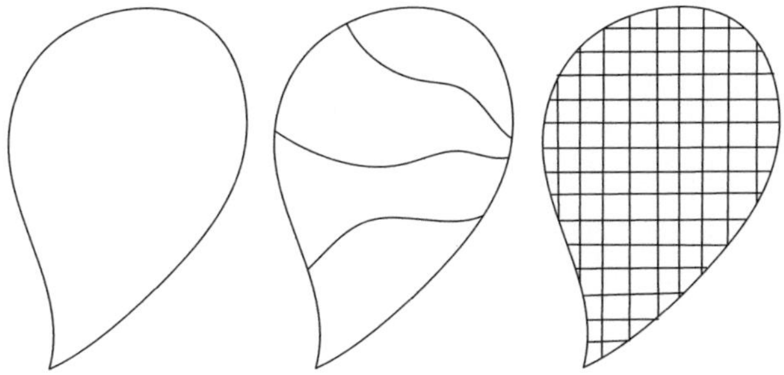

Fig. 1.4 (From *left*) Examples of lumped, semi-distributed, and distributed model

For instance, a model benefits from the average precipitation in a basin. In contrast, in some cases we need a model to be highly sensitive to the location where a variable is collected, generated, or developed through the process of modeling. This case needs a model to be considered as a distributed model within the boundary of the system. A model with the capability of interpolating/extrapolating of rainfall in each location of a watershed is considered as a distributed model. A semi-distributed model is a model which is not sensitive to each location of a region but considers variation within subdistricts of the region and acts like a lumped model within them. A hydrological model which gets different parameters for different subbasins of a river basin but follows the rules and equations of a lumped model within them is known as a semi-distributed model. Figure 1.4 shows schematically the difference between three lumped, distributed, and semi-distributed models. The figure shows the boundaries where the spatial variability of the model's parameters is concerned.

1.3.2 Temporal Complexity

Data-driven models can be either static or dynamic. Static simulation models are based on the current state of a system and assume a constant balance between parameters with no predicted changing. In contrast, dynamic simulation models rely on the detailed assumptions regarding changes in existing parameters by changing the state of a system. A rainfall forecasting model is considered as a dynamic model if its parameters change as it receives new precipitation data in its application time line. In contrast, it may be considered as a static model in case its parameters rely on the historical data with no plan to be changed in the time horizon of its application.

1.4 Model Selection

Model selection is the task of selecting a model from a set of candidate models via number of logical criteria. Undoubtedly among given candidate models of similar accuracy and precision, the simplest model is most likely to be chosen but still they need to be examined closely. Different criteria might be used in the process of selecting a model as follows.

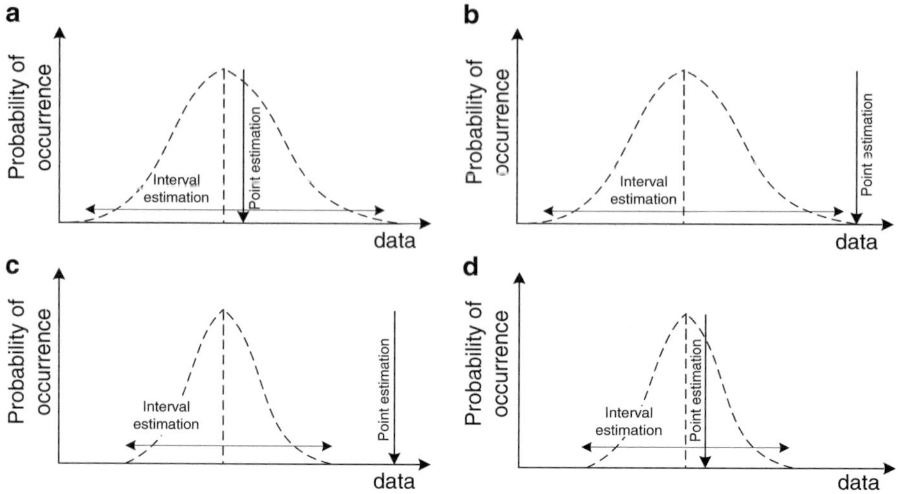

Fig. 1.5 Examples of different models based on precision and accuracy (**a**) Accurate but not precise (**b**) Not accurate and not precise (**c**) Precise but not accurate (**d**) Precise and accurate

1.4.1 Purpose of Application

Obviously, the purpose of modeling, such as research task, designing, and operating, is an important criterion to select a model. Actually, the purpose of modeling determines required complexity, developing time, challenging issues, and run time expected from a model, implicitly.

1.4.2 Accuracy and Precision

Accuracy is determined by a comparison between a real variable and its estimated expected value obtained by a model. An accurate model is the one that provides estimated variables close to those of observed in real world. Precision is defined as the possible or probable estimated interval of a model's estimated variable. A precise model is a model that prevents a wide estimation interval through the process of estimation. While the accuracy of a model is measured by the expected value of its estimated output, the precision of a model is represented by the probable/possible range that the estimated variable of the model might have. The most preferred model is the one which is both accurate and precise.

Figure 1.5 represents different schematic combinations of accuracy and precision of a model. Figure 1.5a represents a model which is accurate but not precise. Figure 1.5b represents a model which is neither precise nor accurate. Figure 1.5c is a model which is precise but not accurate, and finally Fig. 1.5d represents a model which is both accurate and precise.

1.4.3 Availability of Data

Data availability determines the details a model can handle. A very complex model could be completely useless in a case study that the necessary data is not provided in a proper manner. There must be a balance between the details expected from a model with the data which is available for it. Obviously none of them must prevent us modeling a system. It is worth notifying that the availability of data does not only deal with the diversity of applied variables but the records an individual data might have in different times and locations. Some models are unable to process multivariate data and some might need a minimum record of data to perform appropriately. That is why the availability of data must be checked before selecting a model.

1.4.4 Type of Data

Many of the data-driven models have been developed to deal with specific type of data. This might limit their application in the field of water resources and environmental engineering. Models such as ARMA and ARIMA have been developed to deal with the time series data where a model such as static multilayer perceptron has been specifically developed for event-based data and is not able to model the temporal structure of a time series. Models such as probabilistic neural networks usually use discrete data to classify a set of input variables, and models such as fuzzy inference systems can consider descriptive data. The following represents a brief description of different types of data that might help selecting an appropriate model for a specific application.

1.4.4.1 Descriptive and Numerical Data

Data can be *descriptive* (like "high" or "low") or *numerical* (numbers). In the field of water resources and environmental engineering, both descriptive and numerical data could be used. Linguistic variables such as "low consumption," "high quality," and "poor data" are examples of descriptive data in these fields. The use of numerical data is preferred for a wide range of data-driven models and matches better to their techniques. However, developments in the application of fuzzy logic in engineering have spurred the use of descriptive data beside the numerical ones.

1.4.4.2 Discrete and Continuous Data

The numerical data can be *discrete* or *continuous*. Discrete data is counted, and continuous data is measured. Discrete data can only take certain values (Fig. 1.6a).

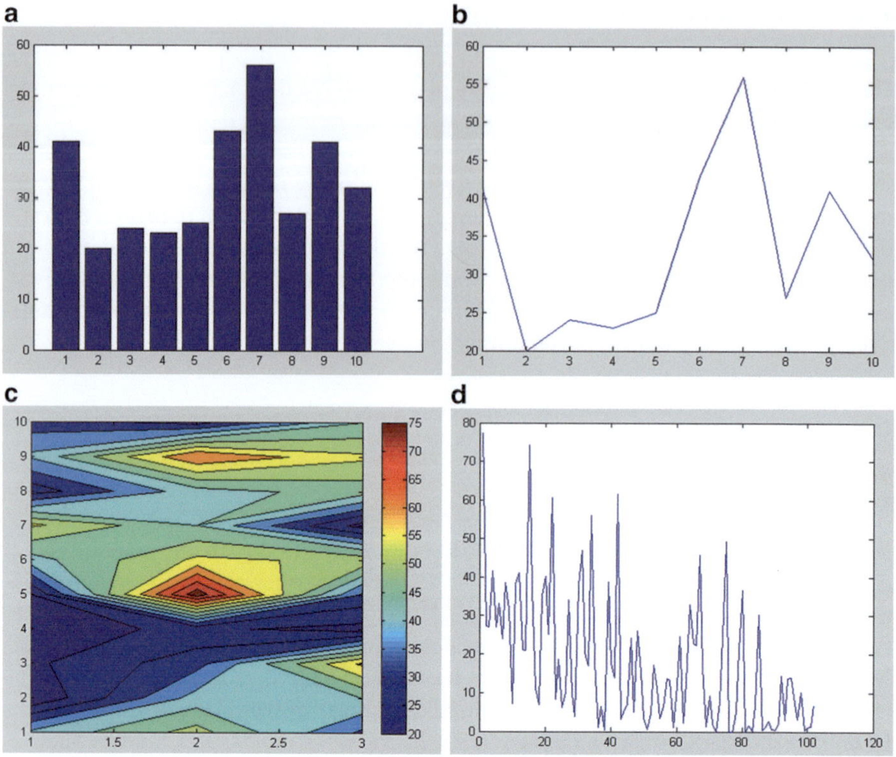

Fig. 1.6 Example of discrete (**a**) Discrete data (**b**) Continuous data (**c**) Spatial data (**d**) Time series

Examples of discrete data are number of floods during water years, duration of drought events, number of rainy days, etc.

Continuous data makes up the rest of numerical data. This is a type of data that is usually associated with some sort of physical measurement (Fig. 1.6b). Examples of continuous data are air temperature, rainfall depth, streamflow, total dissolved solids, etc.

1.4.4.3 Spatial Data

Spatial data also known as geospatial data or geographic information identifies the geographic location of features and boundaries on earth, such as natural or constructed features. Spatial data is usually stored as coordinates and topology and can be mapped (Fig. 1.6c). Examples of spatial data are water quality in different locations of a river or different locations of a water table. Rainfall recorded in different stations of a basin is another example of spatial data.

1.4.4.4 Time Series Data

Time series data are quantities that represent or trace the values taken by a variable over a period such as a month, quarter, or year. Time series data occur wherever the same measurements are recorded on a regular basis (Fig. 1.6d). Examples of time series data are daily and monthly rainfall data, inflow to a reservoir, etc.

1.5 General Approach to Develop a Data-Driven Model

In spite of differences among the details needed for specific data-driven models, they all follow a general approach, which is presented in this section. Different steps of this approach are shown and numbered in Fig. 1.7.

1.5.1 Conceptualization

The term conceptualization refers to the process that brings a concept in mind to the objects and variables which is dealt with by a model. In terms of software engineering, this step tends to a conceptual model. It should be notified that the term "conceptual model" presented here differs from what has been described in Sect. 1.2.

1.5.2 Model Calibration

After the conceptualization stage, data is preprocessed to be ready for the model. Rescaling the input data and re-dimensioning them are two common processes, which are usually followed up through preprocessing. Different input data usually have different orders which need to be rescaled to prevent loosing the value of a certain variable. Furthermore, many data-driven models are sensitive to the dimension of input/output data set especially in problems with limited data which need few parameters to be calibrated in a robust manner. After preprocessing, *calibration*, which in some models is also called *training*, becomes the process of calculating the parameters of a model.

1.5.3 Model Validation

Validation is referred to a process in which a model is evaluated based on a new set of data that are not used in the calibration process. This process represents a microscale of the environment in which a model is applied in a real case. If the result of validation is not satisfactory, a change in the processes of initial steps should be performed to improve the performance of the final model.

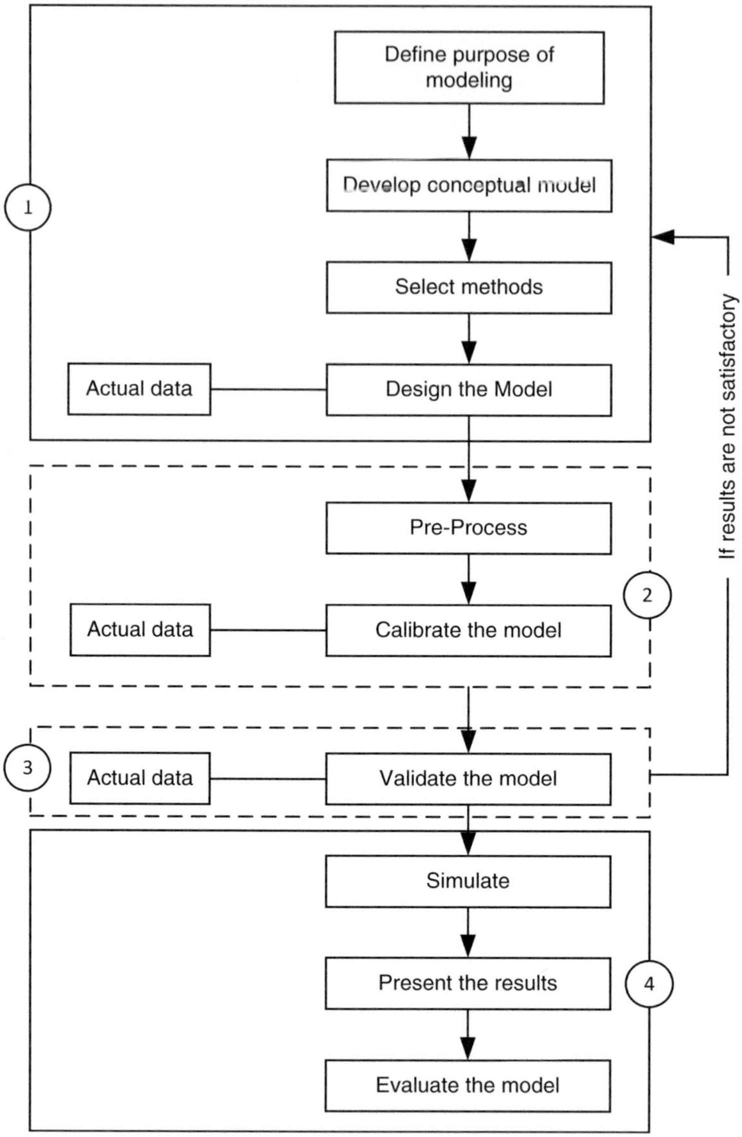

Fig. 1.7 General approach for development of a data-driven model

1.5.4 Presenting the Results

Developed model is used to simulate different states of the system. The results can be shown by different types of tables, charts, and plots through postprocessing. The use of the model is verified through its long-term application in real case studies.

1.6 Beyond Developing a Model

A model is developed to help solving a specific problem; however, in cases that we need to frequently face a generic problem and finding appropriate responses based on the current spatial and temporal state of a system, it is preferred to improve a model to a software. A software is actually a trade version of one or several models in which a designed graphical user interface makes a user-friendly tool for general structured problems.

Decision-making is one of the most significant challenges in the field of water resources and environmental engineering because of either the complexity around a problem or the unpredicted impacts of a decision. This challenge might be a result of multidisciplinary problems, which may put some contrasting objectives in a competition that no compromising is allowed. To overcome this challenge, new technologies have presented powerful tools to increase efficiency and accuracy of decisions and to accelerate the responses in facing with the real-world problems. Decision support systems (DSSs), which are more than a software, are one of the most efficient tools, with distinctive ability approval in the enormous engineering contexts.

Generally, the decision-making procedure includes three main steps, namely, data gathering, recognition of alternatives to solve a specific problem, and, finally, selection of the best alternative. This procedure may be followed by two different approaches. In the first approach, well-known mathematical formulation and decision rules are used algorithmically in different steps of solving a specific problem. Problems that are likely to be solved by this approach are usually called structured problems. These kinds of problems could be possibly solved manually by the use of computer softwares, where no human judgment is needed. In contrast, in some problems, usually called unstructured problems, no decision rules and algorithmic procedure are defined and are dependent considerably to the human judgment to be solved. Decision support systems could be called as the second approach and have been developed to be used in solving the latter.

Decision support systems are set of models developed for the sake of data analysis to help a precise judgment (Little 1970). Little (1970) declared that a decision support system is a simple, robust, controllable, and flexible system containing appropriate subsystems as well as user interfaces. Bonczek et al. (1981) have defined DSSs as computer systems including three interactive components of user *interface*, a *knowledge system*, and a *problem-processing* system. Technology developments have changed slightly the definition of such systems in both holistic and detailed manner. Watkins and McKinney (1995) have defined a DSS as a computer system which uses analytical models to help decision-makers in defining and organizing various alternatives and analyzing their impacts to choose the most appropriate alternatives. In a general definition, the architecture of DSS consists of three components, namely, database, model base, and user interface, as shown in Fig. 1.8.

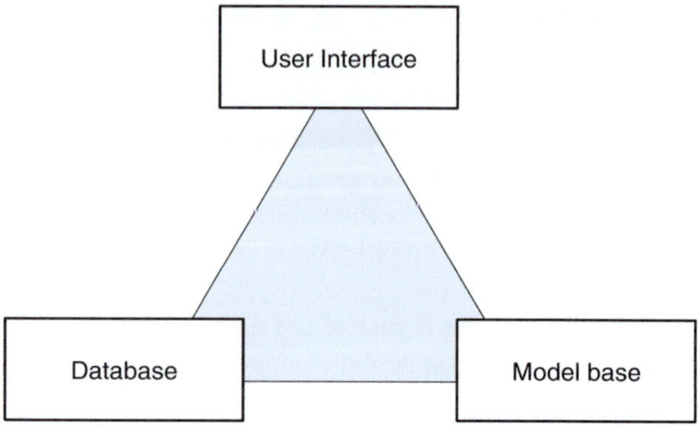

Fig. 1.8 Main components of a decision support system

DSSs usually are developed for certain groups of decision-makers. This needs a specific design such that the decision-maker could define new alternatives and more importantly change an existing alternative to analyze that using the models embedded in the system. Since the delay in responding by the system is considered as an index of inefficiency, an interactive user interface, easy change of input parameters, and quick, understandable, and managed output are considered as characteristics of a DSS.

References

Bonczek RH, Holsapple CW, Whinston A (1981) Foundations of decision support systems. Academic Press, New York

Little JDC (1970) Models and managers: the concept of a decision calculus. Manage Sci 16(8): B466–B485

Watkins DW, Mckinney DC (1995) Recent developments associated with decision support systems in water resources. Rev Geophys 33(Suppl):941–948

Chapter 2
Basic Statistics

Abstract A stochastic variable is a combination of two components of deterministic variable, D, and random variable, ε. While D could be modeled by a range of mathematical models, ε is described by the probability theory using probability distribution function (*pdf*). Regarding the type of a random variable which might be discrete or continuous, it is defined by two types of discrete and continuous *pdfs*. Discrete distribution functions of *Bernoulli*, *binomial*, and *Poisson* are reviewed in this chapter along with the continuous distribution functions of *exponential*, *uniform*, *normal*, and *extreme value*. One of the most applicable fields of distribution functions is *frequency analysis* which is discussed in another section of this chapter. As far as the statistical analysis of real problems is concerned, hypothetical tests are widely used for deciding on either the parameters of one or several populations or the type of a distribution function which better fits the data. The hypothetical tests follow a general approach while that approach should be adapted for specific problems by defining appropriate statistical and critical values. The tests on the statistical parameters of populations are reviewed in this chapter. Furthermore, two famous tests of *chi-square* and *Kolmogorov–Smirnov* are presented to decide on the best distribution function for a specific random variable. Each of the above calculations is supported by the related commands and programs provided in MATLAB.

Keywords Probability distribution function • Frequency analysis • Hypothesis test • Distribution fitting

2.1 Introduction

Stochastic process is a frequently used term in the technical speaking of the field of water resources and environmental engineering. What is exactly called a stochastic process? A stochastic process is a process that deals with both random and deterministic variables. Suppose a stochastic process presented by vector of variables, X. X, is considered as a combination of D and ε. D is a component which could be

S. Araghinejad, *Data-Driven Modeling: Using MATLAB® in Water Resources and Environmental Engineering*, Water Science and Technology Library 67,
DOI 10.1007/978-94-007-7506-0_2, © Springer Science+Business Media Dordrecht 2014

modeled by a range of analytical, conceptual, or data-driven models, known as "*deterministic variable.*" ε is a component which could not be analyzed by any model at all, known as "*random variable.*" It could just be defined by the probability theory with the use of distribution functions.

By the above definition, how is a rainfall data categorized? Is it a random, deterministic, or stochastic variable? Actually, definition of rainfall depends on our knowledge and ability to define the rainfall process and its modeling. Certainly, we cannot consider rainfall as a deterministic variable as it is still beyond our skills to be 100 % sure of what we report as predicted or forecasted rainfall. It could be considered as a random variable when we talk about rainfall estimation for long-term return periods (say a storm with 50-year return period). Meanwhile it could be considered as a stochastic variable in short-term rainfall estimation since application of short-term forecasting tools has made it possible to have estimates of rainfall variable in short lead times, even though the errors of forecasting variable force us to be always uncertain about some fraction of our estimate. Obtained results by tossing a coin or a dice are examples of random variables as we have no control or knowledge to decide about the result before tossing up.

The scope of this chapter is to deal with the random term of a stochastic variable, ε. While the deterministic and stochastic variables are modeled by a range of models, a random variable is not modeled but it is defined and expressed by *probability distribution function*. The uncertainty within a process is actually synonymous to the contribution of D and ε to the process X. The uncertainty increases if the random term (ε) plays a more significant role than D through the process and vice versa. In fact, in a process that the portion of ε is considerably more than the deterministic term, it is useless to try modeling it by data-driven models or any other mathematical type of models. The probability theory becomes the solution for such cases. It should be notified that describing the random term by using the probability theory is also useful to be applied in processes that involve a weal term of random variable. The analyzed random term is representative of the uncertainty of the process and helps risk-based decision-making.

Two major goals are followed up by presenting this chapter. First is to use the basic statistics for preprocessing and postprocessing and in short for better preparation and reporting a set of data. Since the basic concepts of regression models, time series analysis, and statistical neural networks have originated from the basic statistics, the second goal of this chapter is to prepare readers to study the next chapters.

This chapter begins with basic statistical definitions, which is followed by a review on the most famous discrete and continuous probability distribution functions. *Frequency analysis* is another section of this chapter, which is completely related to the application of *pdf*s. The most applicable statistical tests are presented in the next section dealing with two types of tests, those which deal with the statistical parameters of samples and those which deal with the selection of the type of *pdf*s. Finally the chapter ends with a workshop. At a glance, the structure of the contents of this chapter is reviewed in Fig. 2.1.

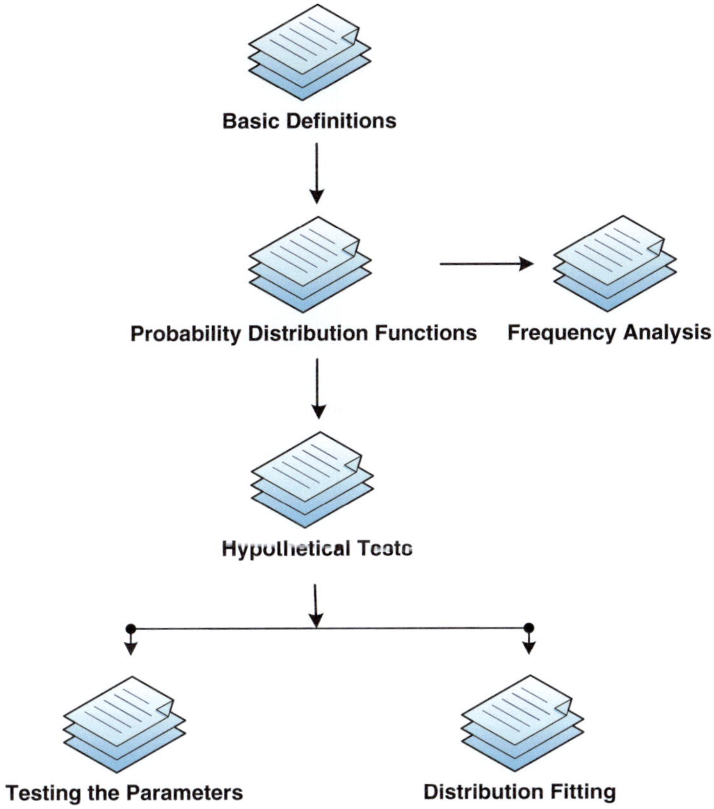

Fig. 2.1 Structure of the contents of this chapter

2.2 Basic Definitions

A random variable consists of a range of values, which are associated with a certain probability of occurrence. For instance, the number of rainy days in November is a discrete random variable, which has the values from 1 to 30, where the days near to the end of November have higher probability of occurrence in comparison to the days of the first half of November. The value of rainfall in November is an example of continuous variables ranging from 0 to 15 mm, where values close to 10 might have higher probability of occurrence. To describe a random variable, it is usual to demonstrate its range of probable values, X, in a horizontal axis and the associated probability of occurrence, $f(x)$, at the vertical axis. This typical figure as shown in Fig. 2.2a is called probability distribution function. In many problems, it is preferred to deal with the probability of a group of variables instead of a specific variable. Therefore, another type of probability distribution function is developed based on the integration of the probabilities associated to the values less than or

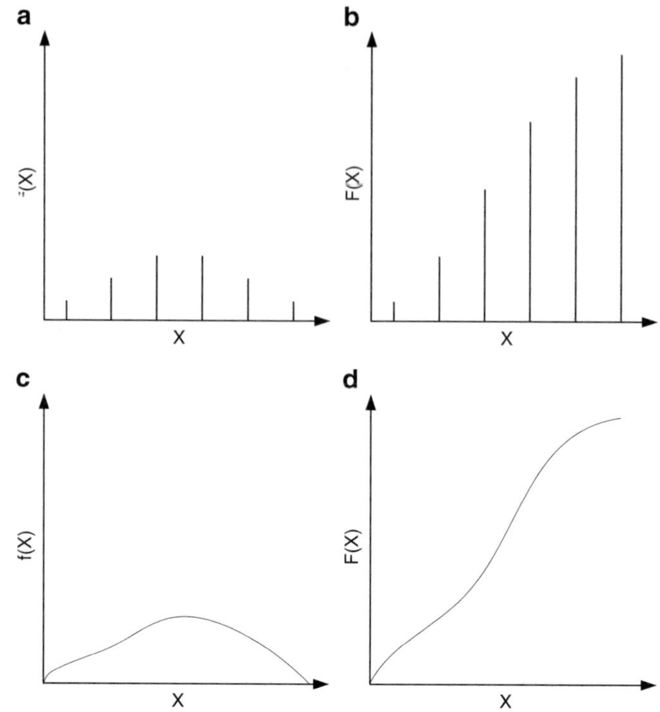

Fig. 2.2 Examples of discrete pdf (**a**) and CDF (**b**) as well as continuous pdf (**c**) and CDF (**d**)

equal to a specific value of x_0. This function is called cumulative distribution function (CDF) (Fig. 2.2b). The discrete form of this function is obtained as

$$F(X = x_0) = \sum_{-\infty}^{x_0} f(X)\Delta x \tag{2.1}$$

where $f(x)$ is the probability distribution function. A continuous CDF for a given value x_0 is calculated by the following relation:

$$F(X = x_0) = \int_{-\infty}^{x_0} f(X)\mathrm{d}x \tag{2.2}$$

It is usual to define a random variable X by the parameters that precisely represent the entire data as well as its probability distribution function. *Statistics* involves the study of data sets like X by describing its statistical parameters. A population includes each element from the set of observations that can be made. The term "*population*" is used in statistics to represent all possible measurements or outcomes that are of interest to us in a particular study. The term "*sample*" refers to a portion of the population that is representative of the population from which it was selected. A sample consists only of observations drawn from the population. A measurable characteristic of a population, such as a mean or standard deviation, is called a *parameter*; but a measurable characteristic of a sample is called a statistic.

The mean and the median are summary measures used to describe the most "typical" value in a set of variables. Mean and median are usually referred as measures of central tendency. The mean of a sample or a population is computed by adding all of the observations and dividing by the number of observations. To find the median, we arrange the observations in order from smallest to largest value. If there is an odd number of an observation, the median is the middle value. If there is an even number of observations, the median is the average of the two middle values.

Let us assume X as a vector of n random continuous numbers, $X = \{x_1, x_2, \ldots, x_n\}$, and $f(X)$ as its probability distribution function. The mean of this variable is obtained as

$$\overline{x} = \int_{-\infty}^{\infty} xf(X)\mathrm{d}x \tag{2.3}$$

Considering a uniform distribution function for X (where all values are associated with the same probability of $1/n$), the mean is obtained by

$$\overline{x} = \frac{1}{n}\sum_{i=1}^{n} x_i \tag{2.4}$$

It should be notified that the mean of a population is denoted by the symbol μ; but the mean of a sample is denoted by the symbol $\overline{x}$, and both are obtained by a similar formulation.

In MATLAB, the following commands are used to calculate the mean of matrix X (MATLAB 2006):

```
M=mean(X,dim)
```

dim = 1 returns the mean of each column, and dim = 2 returns the mean of each row of the matrix.

In case of using a vector instead of a matrix, the command is summarized to

```
M=mean(X)
```

The following command is used to calculate the median of matrix X:

```
M=median (X,dim)
```

Summary measures could be used to describe the amount of variability or spread in a set of data. The most common measures of variability are the range, variance, and standard deviation. The range is the difference between the largest and smallest values in a set of values. In a population, variance is the average squared deviation from the population mean, as defined by the following formula:

$$\sigma^2 = \frac{\sum_{i=1}^{N} (x_i - \overline{x})^2}{N} \tag{2.5}$$

Observations from a sample can be used to estimate the variance of a population. For this purpose, sample variance is defined by slightly different formula and uses a slightly different notation:

$$S^2 = \frac{\sum_{i=1}^{n} (x_i - \overline{x})^2}{n - 1} \tag{2.6}$$

The standard deviation is the square root of the variance. Thus, the standard deviation of a sample is

$$S = \sqrt{\frac{\sum_{i=1}^{n} (x_i - \overline{x})^2}{n - 1}} \tag{2.7}$$

In MATLAB, variance and standard deviation of matrix X are obtained by the following commands:

```
V=var (X,w,dim)
S=std (X,w,dim)
```

$w = 0$ uses $n - 1$, and $w = 1$ uses n in Eq. (2.5). On the other hand, $w = 0$ is used to calculate the variance and standard deviation for a sample, and $w = 1$ is used to calculate the variance and standard deviation for a population.

dim plays the role as described before.

Mode is a number in a vector which has the maximum frequency among the others. In MATLAB it is calculated by the following command:

```
Mo=mode (X,dim)
```

Table 2.1 Streamflow data for Example 2.1

	1	2	3	4	5	6
1	220,158.2	176,491.7	441,836.0	35,907.4	48,308.1	1,243.1
2	222,924.4	170,472.1	440,202.6	36,856.1	48,070.5	1,243.7
3	219,749.8	176,723.9	441,400.6	35,981.0	48,464.2	1,242.8
4	222,537.8	170,488.0	436,882.5	37,278.8	48,493.9	1,243.4
5	222,924.4	170,472.1	440,202.6	36,856.1	48,246.0	1,243.7
6	215,977.3	171,496.7	430,090.5	36,414.4	48,079.4	1,235.0

Range is a static that represents the difference between maximum value and minimum value among a data set and is calculated as follows:

```
R=range (X,dim)
```

Other popular parameters of a random variable are *skewness* and *kurtosis*, which are obtained as

$$\text{Skewness} = \sum_{i=1}^{N} \frac{(x_i - \overline{x})^3}{S^3} \tag{2.8}$$

and

$$\text{Kurtosis} = \sum_{i=1}^{N} \sum \frac{(x_i - \overline{x})^4}{S^4} \tag{2.9}$$

Those statistics are calculated by MATLAB using the following commands:

```
Sk= Skewness (x,flag,dim)
K= Kurtosis (x,flag,dim)
```

flag $= 0$ is used to correct the calculation for a sample from a population. Otherwise flag $= 1$ is used to calculate the above parameters for a population.

Example 2.1: Summary Statistics

For the monthly streamflow data given in Table 2.1, find a summary of statistics.

Solution
Mean of each column is calculated as follows:

```
M=mean(X,1)

M =

   1.0e+005 *

      2.2071   1.7269   4.3844   0.3655   0.4828   0.0124
```

Mean of each row is calculated as follows:

```
M=mean(X,2)

M =

   1.0e+005 *

   1.5399
   1.5329
   1.5393
   1.5282
   1.5332
   1.5055
```

Standard deviation of each row is obtained as follows:

```
S=std (X,0,2)

S =
  1.0e+005 *

     1.6504
     1.6444
     1.6483
     1.6314
     1.6442
     1.6043
```

Finally, *Mode* and *Range* of each column is obtained as

```
Mo=mode (X,1)

Mo =

  1.0e+005 *

    2.2292  1.7047  4.4020  0.3686  0.4807  0.0124
```

and

```
R=range (X,1)

R =

  1.0e+004 *

    0.6947  0.6252  1.1746  0.1371  0.0423  0.0009
```

2.3 Graphical Demonstration of Data

The first step to investigate and demonstrate a set of data is to display them in a graphical form. Two most applicable graphical forms of data are histogram and box plot. Graphical distribution is only used for a quick assessment of data. More details about the data are obtained by calculating parameters and statistics defined before.

2.3.1 Histogram

A histogram is a bar plot of frequency distribution that is organized in intervals or classes. The histogram provides useful information about the data such as central tendency, dispersion, and the general shape of the data distribution. An example of a histogram is shown in Fig. 2.3.

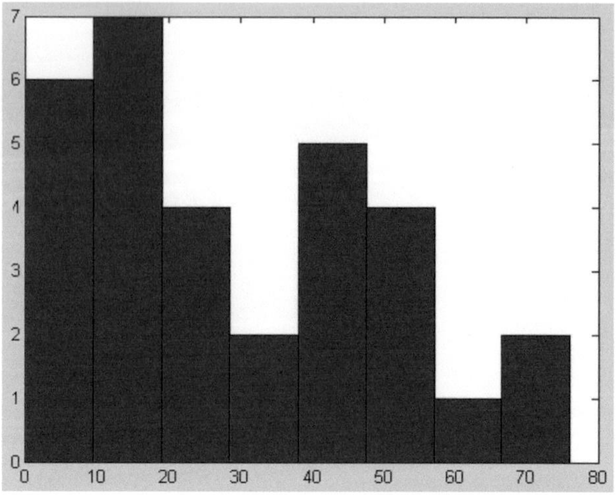

Fig. 2.3 An example of a histogram

A histogram of data, *Y*, is plotted by the following syntax for number of bars equal to *nbins*:

```
Hist(Y,nbins);
```

2.3.2 *Box Plot*

Assume that the elements in a data set are rank ordered from the smallest to the largest. The values that divide a rank-ordered set of elements into 100 equal parts are called percentiles. An element having a percentile rank of P_i would have a greater value than *i* percent of all the elements in the set. Thus, the observation at the 50th percentile would be denoted P_{50}, and it would be greater than 50 % of the observations in the set. An observation at the 50th percentile would correspond to the median value in the set.

Quartiles divide a rank-ordered data set into four equal parts. The values that divide each part are called the first, second, and third quartiles; and they are denoted by Q_1, Q_2, and Q_3, respectively. Q_1 corresponds to P_{25}, Q_2 corresponds to P_{50}, and Q_3 corresponds to P_{75}. Q_2 is the median value in the set.

A box plot, sometimes called a box and whisker plot, is a type of graph used to display patterns of quantitative data. A box plot splits the data set into quartiles. The body of the box plot consists of a *box*, which goes from the first quartile (Q_1) to the third quartile (Q_3). Within the box, a vertical line is drawn at the Q_2, the median of the data set. Two horizontal lines, called whiskers, extend from the front and back of the box (Fig. 2.4). The front whisker goes from Q_1 to the smallest non-outlier in the data set, and the back whisker goes from Q_3 to the largest non-outlier (Trauth 2008).

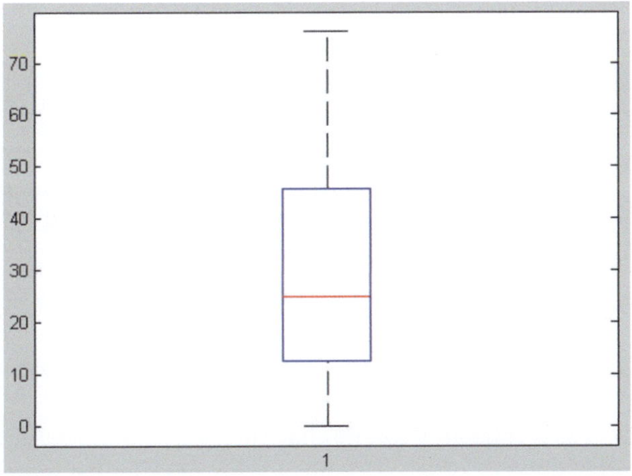

Fig. 2.4 An example of a box plot

A box plot of a vector or a matrix Y is plotted by the following syntax:

```
boxplot(Y);
```

2.4 Probability Distribution Functions

A probability distribution function assigns a probability to each of the probable outcomes of a random variable. In statistics, the empirical distribution function, or empirical CDF, is the cumulative distribution function associated with the empirical measure of the sample. The empirical distribution function estimates the true underlying CDF of the points in the sample. Instead, a theoretical distribution function replaces the empirical measure of samples by a mathematical relation. This enables us to generalize the frequency analysis over a certain and limited sample data. Among the others, the following theoretical distribution functions are the most famous ones, which are frequently used in the field of water resources and environmental engineering:

2.4.1 Binomial Distribution

This distribution gives the discrete probability of x successes out of n trials, with probability p of success in any given trial. The probability distribution function of binomial distribution is

$$f(x) = \binom{n}{x} p^x q^{n-x} \tag{2.10}$$

where

$$\binom{n}{x} = \frac{n!}{x!(n-x)!} \tag{2.11}$$

In case of just one trial ($n = 1$), binomial distribution changes to the well-known *Bernoulli* distribution. Bernoulli distribution could be considered as the simplest theoretical distribution function.

Example 2.2: Binomial Distribution

The probability of rainfall at each day of a month is 1/12. Calculate the probability of observing only 4 rainy days in that month.

Solution

This is an example of a binomial distribution where n is the number of days in the month, p is the probability of raining, and x is the number of rainy days.

$$f(4) = \binom{30}{4} {}^{1}\!/_{12}{}^{4}\ {}^{11}\!/_{12}{}^{26} = 0.138 = 13.8\ \%$$

The solution can also be obtained by MATLAB using the following syntax, where $x =$ number of successes, $n =$ number of trials, and $p =$ probability of success:

```
y=binopdf (x,n,p)
```

The above syntax for this example changes to

```
y=binopdf (4,30,1/12)
```

Furthermore, the CDF values of binomial distribution can be obtained by the following syntax:

```
y=binocdf (x,n,p)
```

Using the graphical user interface for statistical distributions in MATLAB, which is run by disttool, the following pdf and CDF for binomial distribution are plotted, which are associated with Example 5.2 (Fig. 2.5).

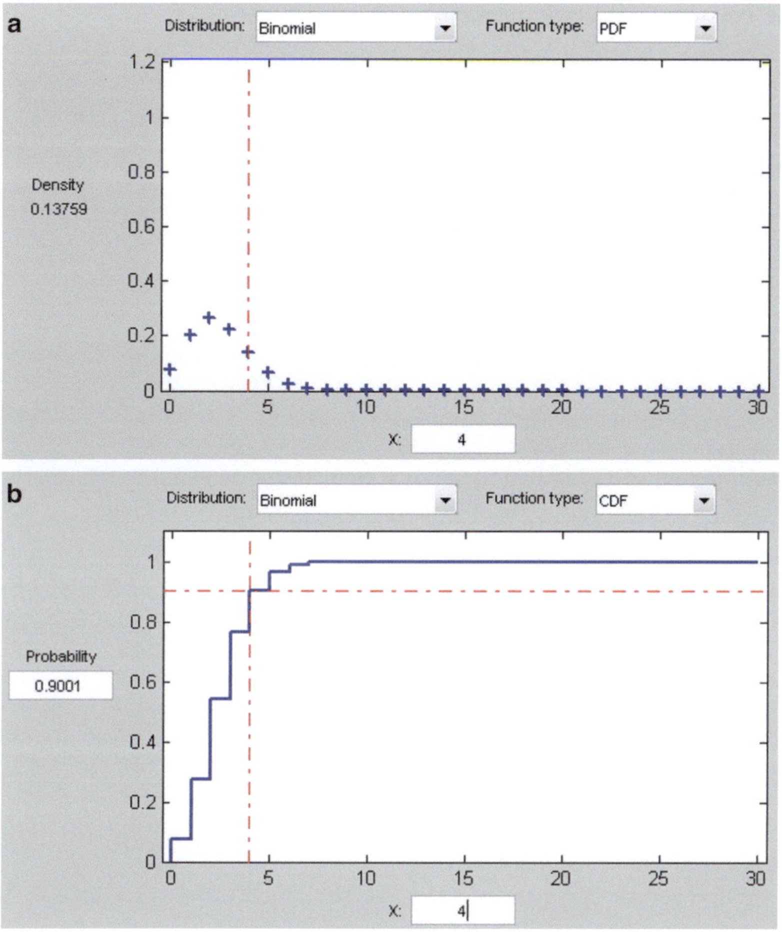

Fig. 2.5 pdf (**a**) and CDF (**b**) of binomial distribution for Example 2.2

2.4.2 Poisson Distribution Function

When the number of trials increases and the probability of success (or failures in most cases) decreases, the binomial distribution function approaches the Poisson distribution. Poisson distribution usually describes the errors in a life time of a system. Considering the system as the natural environment, the errors are in fact droughts, floods, failures of the water structures, pollution hazard, etc. In those cases p is actually the probability of the occurrence of the phenomenon, n is the number of observations, and the parameter of the Poisson distribution is defined as $\lambda = np$.

The Poisson distribution function is presented as

$$f(x) = \frac{e^{-\lambda}\lambda^x}{x!} \tag{2.12}$$

and its cumulative distribution function is

$$F(x) = \sum_{i-0}^{x} \frac{e^{-\lambda}\lambda^i}{x!} \tag{2.13}$$

Example 2.3: Poisson Distribution

Return period of a flood is 10 years. Calculate the probability of experiencing such flood two times during the 4-year construction time of a bridge.

Solution
The parameter of the Poisson distribution is $\lambda = np = 4 \times 0.1 = 0.4$. The number of failures, x, is equal to 2. This problem can be solved by the following syntax:

```
y=poisspdf (x,landa)
```

It is changed for this specific example as

```
y=poisspdf (2,0.4)
```

which results in almost probability of 0.054. By changing x to 1, the probability is obtained as 3 %.

To calculate CDF values of Poisson distribution, the following syntax is used:

```
y=poisscdf (x,landa)
```

Using `disttool`, the following *pdf* and *CDF* of Poisson distribution are plotted which are associated with the above problem (Fig. 2.6).

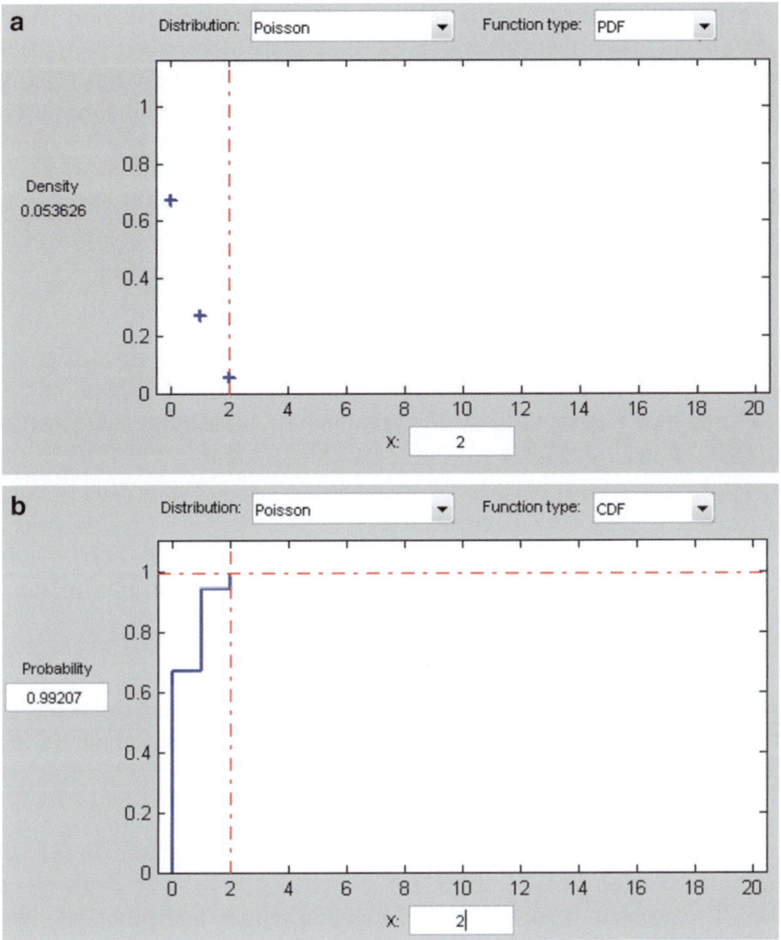

Fig. 2.6 pdf (**a**) and CDF (**b**) of Poisson distribution for Example 2.3

2.4.3 Exponential Distribution Function

How much time will pass before a flood hazard occurs in a given region? How long will it take before an agricultural farm receives a rainfall? How long will water resources system work without breaking down? Questions like these are often answered in a probabilistic manner using the exponential distribution. All of these questions concern the time we need to wait before a given event occurs. If this waiting time is unknown, it is often appropriate to think of it as a random variable having an exponential distribution. Roughly speaking, the time we need to wait before an event occurs has an exponential distribution if the probability that the event occurs during a certain time interval is proportional to the length of that time interval.

The exponential distribution is related to the Poisson distribution. When the event can occur more than once and the time elapsed between two successive occurrences is exponentially distributed and independent of previous occurrences, the number of occurrences of the event within a given unit of time has a Poisson distribution. While Poisson distribution function is used to define the number of failures (or errors) in a system, exponential distribution function is used to define the time interval between two failures. Time is a continuous variable, so the exponential distribution function becomes a continuous distribution function.

The exponential probability distribution function is calculated as

$$f(x) = \frac{1}{\mu} e^{-\frac{x}{\mu}} \tag{2.14}$$

where x is the time between two events, μ is the average waiting time between two events. CDF of an exponential distribution is calculated as

$$F(x) = 1 - e^{-\frac{x}{\mu}} \tag{2.15}$$

Example 2.4: Exponential Distribution

During the construction of a bridge, calculate the risk of experiencing a flood with the return period of 5 years, if the construction time lasts about 18 months. What do you suggest to decrease the risk?

Solution

The parameter of exponential distribution, *mu*, for this problem is the average waiting time between two floods, which is actually 5 years. x is 18 months or 1.5 years. The problem is solved by the following syntax. It should be notified that any x less than or equal to 1.5 years might be considered equal to destruction of the bridge. Therefore, CDF is used to find the probability of x less than or equal to 1.5. CDF of exponential distribution is calculated as

```
Y = expcdf(x,mu)
```

which is changed to

```
Y = expcdf(1.5,5)
```

It results in $y = 0.26$ as the risk of failure. To decrease this risk, we can either reduce the construction time or increase the return period of design flood by strengthening the construction site.

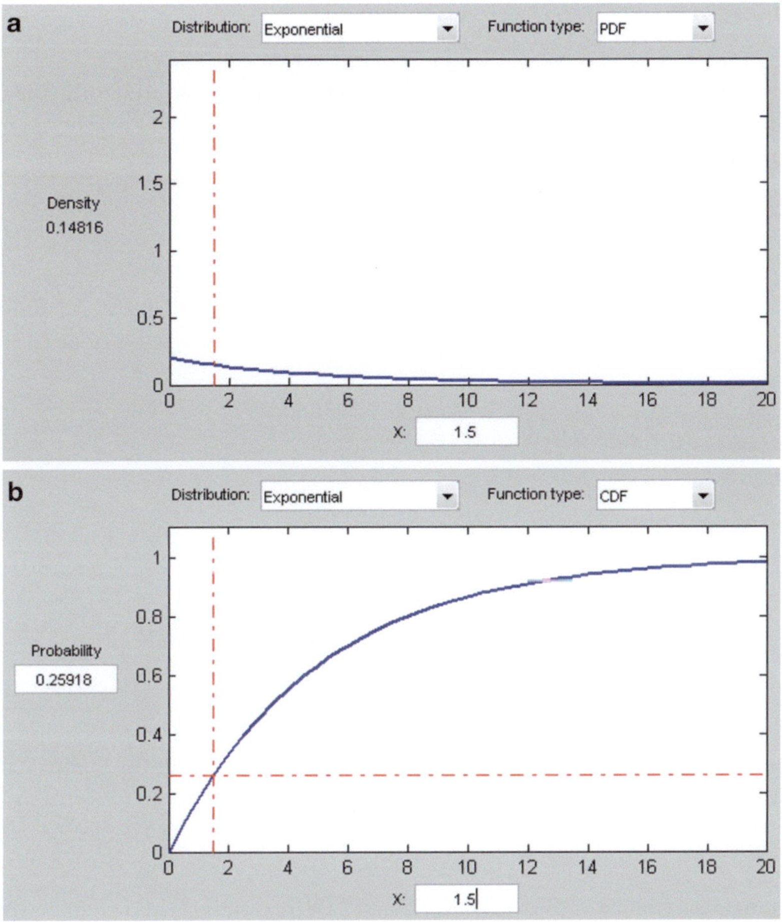

Fig. 2.7 pdf (**a**) and CDF (**b**) of Poisson distribution for Example 2.4

The probability distribution function of exponential distribution can also be calculated by

```
Y = exppdf(x,mu)
```

Using disttool, the following pdf and CDF of exponential distribution are plotted for Example 2.4 (Fig. 2.7).

2.4.4 Uniform Distribution Function

A very simplified distribution function considers a uniform probability for each random variable. If the random variables vary from minimum a to maximum b, then the uniform probability will be as follows:

$$f(x) = \frac{1}{b-a} \quad \text{for} \quad a \le x \le b \tag{2.16}$$

The set of random variables of uniformly distributed will have the following mean and variance:

$$E(x) = \frac{a+b}{2} \tag{2.17}$$

$$\mathrm{var}(x) = \frac{(a-b)^2}{12} \tag{2.18}$$

Example 2.5: Uniform Distribution

Consider a basin that experiences snow precipitation during the winter with the height between 5 and 10 mm. If the snow precipitation follows a uniform distribution, calculate mean and variance of the snow in the basin.

Solution
The mean and variance of the snow are 7.5 and 2.08, respectively, as calculated below:

$$E(x) = \frac{10+5}{2} = 7.5$$

and

$$\mathrm{var}(x) = \frac{(10-5)^2}{12} = 2.08$$

2.4.5 Normal Distribution Function

A normal distribution is often used as a first approximation to describe random variables that cluster around a single mean value. The normal distribution is considered the most prominent probability distribution in statistics. A normally distributed variable has a symmetric distribution about its mean. The normal distribution is defined by the following relation:

$$f(x) = \frac{1}{\sigma\sqrt{2\pi}} \exp\left(\frac{-1}{2}\left(\frac{x-\mu}{\sigma}\right)^2\right) \tag{2.19}$$

and its cumulative distribution function is

$$F(x) = \frac{1}{\sigma\sqrt{2\pi}} \int_{-\infty}^{x} \exp\left(\frac{-1}{2}\left(\frac{y-\mu}{\sigma}\right)^2\right) dy \tag{2.20}$$

The parameters of a normal distribution function are actually the mean and standard deviation of data, μ and σ. There are numerous normal distribution functions due to the change of μ and σ. Among the others, the normal distribution with $\mu = 0$ and $\sigma = 1$ is called normal standard distribution, which is represented as

$$f(x) = \frac{1}{\sqrt{2\pi}} \exp\left(-\frac{z^2}{2}\right) \tag{2.21}$$

where z is a set of random variables with $\mu = 0$ and $\sigma = 1$. The importance of this distribution is that every normal variable can be represented by the normal standard distribution if it is standardized by the following transformation:

$$Z = \frac{X-\mu}{\sigma} \tag{2.22}$$

where X is the original data with mean and standard deviation of μ and σ, respectively. Z is the normal standard data with mean and standard deviation equal to 0 and 1, respectively.

It should be notified that quantities that grow exponentially, such as maximum river discharges, are often skewed to the right and hence may be better described by other distributions, such as the lognormal distribution or the Pareto distribution.

Example 2.6: Normal Distribution Function

For a river with the annual data given in Table 2.2, find the answers of the following questions. The annual average of the river is 1,115 and its standard deviation is 255 million cubic meters.

1. What is the probability of experiencing an annual streamflow less than 1,000 MCM?

Table 2.2 Annual streamflow data for Example 2.6

Year	Streamflow data
2001	1,300
2002	1,600
2003	1,500
2004	1,100
2005	900
2006	770
2007	950
2008	850
2009	980
2010	1,230
2011	1,050
2012	1,150

2. What is the chance of experiencing a streamflow between 1,200 and 900?
3. What is the probability of experiencing an extreme streamflow volume more than 1,500 or less than 700 MCM? (Table 2.2)

Solution

First, the mean and standard deviation of the data are calculated as follows:

```
mu=mean(X)

mu =

          1115
```

and

```
sigma=std(X,0)

sigma =

   255.3963
```

Answer 1.

```
y=normcdf(1000, mu, sigma)

y =

     0.3191
```

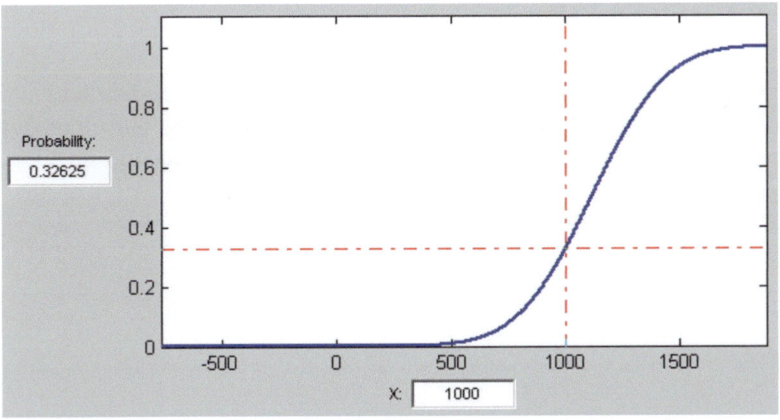

Fig. 2.8 CDF of normal distribution for Example 2.6

Answer 2.

```
y=normcdf (1200, mu, sigma)-normcdf(900,mu,sigma)

y=

     0.4463
```

Answer 3.

```
y=(1-normcdf   (1500, mu,   sigma))+normcdf(700,mu,
sigma)

y=

  0.1025
```

The normal distribution associated to Example 2.6 is shown in Fig. 2.8.

2.4.6 Lognormal Distribution Function

Lognormal distribution gives the opportunity to fit the normal distribution on the data, which are not originally normal. The simple idea here is to transform the original data to normal variables by applying logarithm function on them. The probability distribution function and cumulative distribution function of this distribution are obtained as follows:

$$f(x) = \frac{1}{\sigma\sqrt{2\pi x}} \exp\left(\frac{-1}{2}\left(\frac{\ln x - \mu}{\sigma}\right)^2\right) \tag{2.23}$$

and

$$F(x) = \frac{1}{\sigma\sqrt{2\pi}} \int_{-\infty}^{x} \frac{1}{y} \exp\left(\frac{-1}{2}\left(\frac{\ln(y) - \mu}{\sigma}\right)^2\right) \mathrm{d}y \tag{2.24}$$

2.5 Frequency Analysis

Frequency analysis refers to a process that tries to find the probability of occurrence of a specific variable among the others, by the use of historical records. It also refers to the process which determines a specific value of a variable, which is associated to a predetermined probability. Frequency analysis has a widespread field of application in designing of water resources and environmental variables. A relation that is generally used for frequency analysis is

$$x_T = \overline{x} \pm K S_X \tag{2.25}$$

where $\overline{x}$ and S_X = average and standard deviation of sample data; x_T = specific value of the original data with T return period; and K = a parameter that is a function of T and the probability distribution function that better fits the data.

Example 2.7: Frequency Analysis

Statistical analysis demonstrates that the average and standard deviation of the maximum annual discharge in a river is 16,421 and 1,352 cubic meters per second (cms), respectively. Calculate the maximum annual discharge associated with the 1,000-year return period.

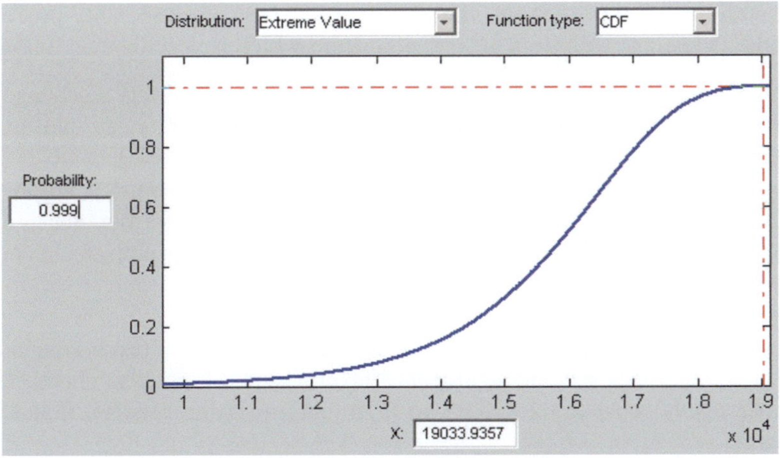

Fig. 2.9 Distribution used for Example 2.7

Solution

A discharge with 1,000-year return period is a discharge which the probability of occurrence such a discharge and greater than that each year is less than 0.001. On the other hand, the cumulative probability of a discharge greater than a 1,000-year discharge is more than 0.999. Using the extreme value distribution (which is usually used for extreme data), the answer will be a discharge of 19,034 cm as shown in Fig. 2.9.

2.6 Hypothetical Tests

Many decisions in the field of engineering involve a choice from two alternatives. For example, the major considerations between two development plans involve a series of decisions as to whether to avoid construction of a dam or change its operational rule. In this case sample information could be used to help decision-making process. The information may involve samples of water quality before and after developing the new dam. The observation of such samples is viewed as tests. These tests are analyzed to see whether the assumption of "constant average water quality" before and after the development plan is correct. If the average value of water quality changes statistically before and after development of the dam, it is concluded that the development plan impacts the water quality. The assumption of "constant average water quality" is refereed as hypothesis. The process which helps decision-makers to decide based on the "assumption" and "tests" is called "hypothesis testing."

Regardless of the type of the decision-making problem, the process of "hypothesis testing" follows a general algorithm which is described as follows:

- Chose a null hypothesis, H_0. Null hypothesis usually represents no change from the natural variation in the parameter of the test. For example, the null hypothesis is "no change in the average of water quality of the river."
- Chose an alternative hypothesis, H_1. The alternative hypothesis is any hypothesis that is different from H_0. For example, the alternative hypothesis in the above example is "the average of water quality of the river worsens in comparison to its natural value."
- Choose a significance level, α.
  α is the error of rejecting H_0 when it is true. Every statistical test is concluded in association with a specific significance level. Often, researchers choose significance levels equal to 0.01, 0.05, or 0.10, but any value between 0 and 1 can be used.
- Calculate a "test statistic" using sample data.
  The "test statistic" is changed by the type of the problem. It will be presented for typical problems in the next expressions.
- Calculate "critical values" obtained from the well-known statistical tables.
  The "critical value" is usually a specific value of a probability distribution function.
- Compare computed test statistic with the critical values.
- If the test statistic falls within the critical limits, then accept H_0, otherwise, reject H_0 and accept H_1.

2.6.1 Testing the Parameters

A statistical hypothesis is usually an assumption about a population parameter. This assumption may or may not be true. Hypothesis testing refers to the formal procedures used by statisticians to accept or reject statistical hypotheses. The best way to determine whether a statistical hypothesis is true would be to examine the entire population. Since that is often impractical, researchers typically examine a random sample from the population. If sample data are not consistent with the statistical hypothesis, the hypothesis is rejected.

2.6.1.1 Testing the Population Mean

In real problems, there might be different questions regarding the average of one or two data sets. In case of dealing with a single sample of a statistical set (population) with a specific number of members, we might test whether the average of the population set is equal to a known value or whether the average of a set is equal to the mathematical average of the sample of data. The procedure for these tests is based on the general procedure which was described before:

- The null hypothesis is $H_0 : \mu = \mu_0$.
- The alternative hypothesis is $H_1 : \mu \neq \mu_0$ or $H_1 : \mu > \mu_0$ or $H_1 : \mu < \mu_0$.

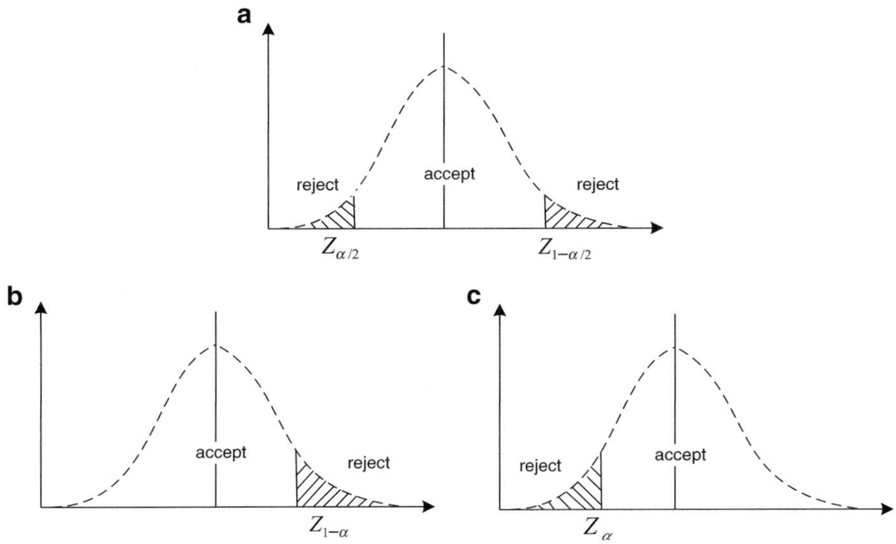

Fig. 2.10 "Reject" and "accept" regions for testing the population mean of single sample in case of known variance (**a**) Two-tailed hypothesis (**b**) One-tailed hypothesis (*right*) (**c**) One-tailed hypothesis (*left*)

- The test statistic is $z = \frac{\overline{X} - \mu_0}{\sigma/\sqrt{n}}$.
- The critical value is obtained from the table of normal standard distribution function.

The following command performs hypothesis test of the null hypothesis that data in the vector X is a random sample from a normal distribution with mean m and standard deviation *sigma*, against the alternative that the mean is not m. The result of the test is returned in h. $h = 1$ indicates a rejection of the null hypothesis at the *alpha%* significance level. $h = 0$ indicates a failure to reject the null hypothesis at the *alpha%* significance level.

```
h= ztest(X,m,sigma,alpha,tail,dim)
```

"dim" is used to perform the test for a column (=1) or a row (=2) of a specific matrix.

The alternative hypothesis may be two-tailed as

$$H_1 : \mu \neq \mu_0$$

In this case the *tail* in the above syntax is assigned as *both*. This case refers to the *reject* and *accept* regions as demonstrated in Fig. 2.10a.

The alternative hypothesis may be one-tailed as

$$H_1 : \mu > \mu_0$$

Table 2.3 Rainfall data for Example 2.9

No.	Rainfall (mm)
1	210
2	213
3	280
4	250
5	230
6	200
7	215
8	2,118

or

$$H_1 : \mu < \mu_0$$

In this case, the *tail* in the above syntax is assigned as "*right*" *or* "*left*" for $\mu > \mu_0$ and $\mu < \mu_0$, respectively. This case refers to the "*reject*" and "*accept*" regions as demonstrated in Fig. 2.10b, c.

X can be a matrix or an n-dimensional array. For matrices, the above test performs separate tests along each column of X and returns a vector of results if *dim* is 1.

If the variance of the population set is an unknown value, normal distribution is replaced by t distribution, test statistic is replaced by $\frac{\overline{X}-\mu_0}{\hat{S}/\sqrt{n}}$, and the following syntax is used to run the test:

```
h= ttest(X,m,alpha,tail,dim)
```

Example 2.8: Testing the Mean for One Sample

Long-term mean rainfall in a region has been 230 mm based on the 50-year recorded data. The recorded rainfall in the last 8 years is reported in Table 2.3. Can we conclude that the mean rainfall value has decreased in recent years?

We use the following command to test the mean for one sample:

```
h= ttest(X,230,0.05,'left')
```

which results in $h = 0$. It means the assumption that the average of data is 230 mm could not be rejected considering 5 % significance level.

Table 2.4 The TDS data of Example 2.10

No.	X	Y
1	130	141
2	135	132
3	140	145
4	128	120
5	145	133
6	130	122
7	125	128
8	127	129
9	129	135
10	130	133

In case of dealing with a two-sample statistical set (population), we might test whether the average of two population sets is equal to a known value, or whether the average of two sets is equal to the mathematical average of two samples. The procedure for these tests is as follows:

The null hypothesis is $H_0 : \mu_1 = \mu_2$.
The alternative hypothesis is $H_1 : \mu_1 \neq \mu_2$ or $H_1 : \mu_1 > \mu_2$ or $H_1 : \mu_1 < \mu_2$.
The test statistic is $\frac{(\overline{X}_1 - \overline{X}_2)}{\sqrt{(\sigma_1{}^2/n_1) + (\sigma_2{}^2/n_2)}}$.

If the variance of the population sets is an unknown value, normal distribution is replaced by t distribution, test statistic is replaced by $\frac{(\overline{X}_1 - \overline{X}_2)}{\sqrt{(S_1^2/n_1) + (S_2^2/n_2)}}$, and the following syntax is used to run the test:

```
h= ttest2(x,y,alpha,tail)
```

Example 2.9: Testing the Mean for Two Samples

Table 2.4 shows two sets of measured total dissolved solids in a river station before development of a surface reservoir upstream of the station (X) and after that (Y). Test the possible increasing of TDS after the development.

Solution
We use the following command to test the mean for two samples:

```
h= ttest2(X,Y,0.05,'both')
```

Table 2.5 Null hypothesis, assumptions, and test statistic for testing the mean

Null hypothesis	Assumption	Test statistic	MATLAB command
$\mu = \mu_0$	σ^2 is known	$\frac{\overline{X}-\mu_0}{\sigma/\sqrt{n}}$	ztest
$\mu = \mu_0$	σ^2 is unknown	$\frac{\overline{X}-\mu_0}{S/\sqrt{n}}$	ttest
$\mu_1 = \mu_2$	${\sigma_1}^2$, ${\sigma_2}^2$ are known	$\frac{(\overline{X}_1-\overline{X}_2)}{\sqrt{({\sigma_1}^2/n_1)+({\sigma_2}^2/n_2)}}$	ztest2
$\mu_1 = \mu_2$	${\sigma_1}^2$ and ${\sigma_2}^2$ are unknown	$\frac{(\overline{X}_1-\overline{X}_2)}{\sqrt{(S_1^2/n_1)+(S_2^2/n_2)}}$	ttest2

It results in $h = 0.0$, which means we cannot reject the null hypothesis, which means no significant difference exists between the mean of two data sets.

Table 2.5 shows the summary of the tests used for testing the mean.

2.6.1.2 Testing the Population Variance

The following command performs a test by the null hypothesis that data in the vector X are a random sample with standard deviation σ_0, against the alternative that the variance is not σ_0. The result of the test is returned in h. $h = 1$ indicates a rejection of the null hypothesis at the *alpha*% significance level. $h = 0$ indicates a failure to reject the null hypothesis at the *alpha*% significance level.

```
h= vartest(X,V,alpha,tail)
```

The alternative hypothesis may be two-tailed as

$$H_1 : \sigma \neq \sigma_0$$

In this case, the *tail* in the above syntax is assigned as *both*. This case refers to the *reject* and *accept* areas as demonstrated in Fig. 2.11a.

The alternative hypothesis may be one-tailed as

$$H_1 : \sigma^2 > {\sigma_0}^2$$

or

$$H_1 : \sigma^2 < {\sigma_0}^2$$

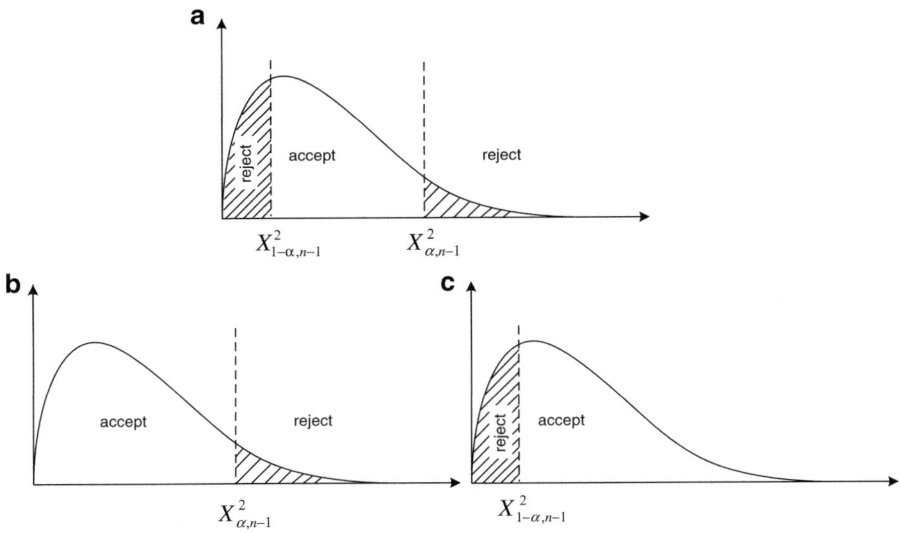

Fig. 2.11 "Reject" and "accept" regions for testing the population variance of a single sample (**a**) Two-tailed hypothesis (**b**) One-tailed hypothesis (*right*) (**c**) One-tailed hypothesis (*left*)

Table 2.6 Null hypothesis, assumptions, and test statistic for testing the variance

Null hypothesis	Assumption	Test statistic	
$\sigma^2 = {\sigma_0}^2$	μ is known	$\frac{(n-1)S^2}{\sigma_0^2}$	vartest
${\sigma_1}^2 = {\sigma_2}^2$	μ_1 *and* μ_2 are known	$F_0 = \frac{S_1^2}{S_2^2}$ *or* $F_0 = \frac{S_2^2}{S_1^2}$	vartest2

In this case, the *tail* in the above syntax is assigned as *right or left* for $\sigma^2 > {\sigma_0}^2$ and $\sigma^2 > \sigma_0^2$, respectively. These cases refer to the reject and accept regions as demonstrated in Fig. 2.11b, c.

In case of dealing with a two-sample statistical set (population), the following test is performed to test whether the variance of two *X* and *Y* sets is equal. Table 2.6 shows the null hypothesis, assumption, and test statistic for testing the population variance.

```
h= vartest2(X,Y,alpha,tail)
```

Example 2.10: Testing the Variance for Two Samples

Test whether the variance of *X* and *Y* data sets presented in Table 2.4 is equal.

Solution
The following command is used:

```
h= vartest2(X,Y,0.05,'both')
```

It results in $h = 1$. It means that in 5 % significance level, the hypothesis of equivalent variances is rejected.

2.6.2 Distribution Fitting

Distribution fitting is referred to as a process of finding a theoretical distribution that better fits the experimental distribution of data. The benefit of using a theoretical distribution instead of an experimental one is to apply it through the frequency analysis and similar statistical modeling. Two famous tests are used for this purpose, namely, *chi-square* test and *Kolmogorov–Smirnov* test. Chi-square test follows the general approach of statistical test as described before. The idea of both tests is to calculate the difference between theoretical and experimental frequency of different classes of data. This difference is used as the test statistic. This statistic is then compared with a critical value which is obtained for each test, specifically.

The test statistic of chi-square test is calculated as

$$\widehat{\chi}^2 = \sum_{k=1}^{K} \frac{(y_k - \hat{y}_k)^2}{\hat{y}_k} \tag{2.26}$$

where K = number of classes where the data are grouped in, y_k = experimental frequency of the data in kth class, and $\hat{y}_k$ = theoretical frequency of the data in kth class obtained from the theoretical *pdf* under the test. The test statistic is actually a chi-square random variable which is compared with the critical value of $\widehat{\chi}^2$ from its theoretical distribution.

The test statistic for *Kolmogorov–Smirnov* test involves

$$D = \max\left(\left|F(x) - \hat{F}(x)\right|\right. \tag{2.27}$$

where $F(x)$ and $\hat{F}(x)$ are the cumulative probability of data (X) classes from the experimental and theoretical cumulative distribution function, respectively. The

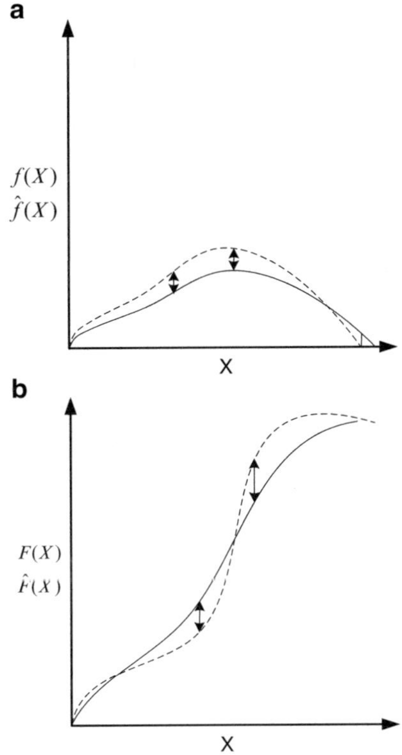

Fig. 2.12 The difference between an experimental and a theoretical distribution as the basic idea of chi-square (**a**) and Kolmogorov–Smirnov (**b**) tests. In the figure, the *dashed line* shows the experimental distribution and the other line shows the theoretical distribution

basic idea supporting both tests is to calculate the difference between the experimental distribution of data and the theoretical distribution function of the test. The chi-square test deals with the *pdf* of distributions, where the Kolmogorov–Smirnov uses the cumulative distribution functions as demonstrated in Fig. 2.12.

Example 2.11: Distribution Fitting

Test whether the drought index observed in a region follows a standard normal distribution (Table 2.7).

Solution

The following command performs a Kolmogorov–Smirnov test to compare the values in the data vector *X* with a standard normal distribution. The null hypothesis for the Kolmogorov–Smirnov test is that *X* has a standard normal distribution. The alternative hypothesis that *X* does not have that distribution. The result h is 1 if the hypothesis that *X* has a standard normal distribution is rejected. *H* is 0 if that hypothesis cannot be rejected. The hypothesis is rejected if the test is significant at the 5 % level.

Table 2.7 Drought index data for Example 2.11

No.	X	No.	X
1	−3.2	11	3.2
2	0.9	12	1.2
3	1.2	13	1.4
4	−1.5	14	−2.1
5	1.5	15	−0.5
6	1.24	16	0.2
7	1.3	17	−0.3
8	1.5	18	0.5
9	−0.5	19	−2.1
10	−1	20	−1.3

```
h= kstest(X)
```

The following command performs a chi-square goodness-of-fit test of the data in the vector X against the normal distribution with mean and variance estimated from X:

```
h= chi2gof(X)
```

The result obtained by both test is $h = 0$.

2.7 Summary

A stochastic variable is considered as a combination of two components: deterministic variable, D, and random variable, ε. While D could be modeled by a range of mathematical models, ε is described by a probabilistic term, known as probability distribution function (*pdf*). Regarding the type of a random variable which might be discrete or continuous, it is defined by two types of discrete and continuous *pdf*s. Discrete distribution functions of *Bernoulli*, *binomial*, and *Poisson* have been reviewed along with the continuous distribution functions of *exponential*, *uniform*, and *normal* distributions.

Bernoulli and binomial distributions are used to define the number of successes (or failures) among limited number of trials. Poisson distribution is used to define number of failures when number of trials increases to a number that could not be handled by binomial distribution. The waiting time between two events is described by exponential distribution. Uniform distribution is used when the probability of random variables is considered uniform. A normal distribution is often used as a first approximation to describe real-valued random variables that cluster around a single mean value.

One of the most applicable fields of distribution functions is frequency analysis. Frequency analysis refers to a process that tries to find the probability of occurrence of a specific variable among the other values of that variable, via the usage of historical records. It also refers to the process which determines a specific value of a variable associated to a predetermined probability.

As far as the statistical analysis of real problems is concerned, hypothetical tests are widely used for deciding on either the parameters of one or several populations or the type of a distribution function that better fits the data. The hypothetical tests follow a general approach while that approach should be adapted for specific problems by defining appropriate statistic and critical values. The tests on the average and variance of one and two populations were reviewed in the chapter. Furthermore, two famous tests of *chi-square* and *Kolmogorov–Smirnov* were presented to decide on the best distribution function for a specific random variable.

References

The Mathworks (2006) Statistics toolbox users guide-for the use with MATLAB. The Mathworks, Natick

Trauth MH (2008) MATLAB recipes for earth sciences. Springer, Berlin/New York

Chapter 3
Regression-Based Models

Abstract Regression analysis aims to study the relationship between one variable, usually called the dependent variable, and several other variables, often called the independent variables. These models are among the most popular data-driven models for their easy application and very well-known techniques. Regression models range from linear to nonlinear and parametric to nonparametric models. In the field of water resources and environmental engineering, regression analysis is widely used for prediction, forecasting, estimation of missing data, and, in general, interpolation and extrapolation of data. This chapter presents models for point and interval estimation of dependent variables using different regression methods. Multiple linear regression model, conventional nonlinear regression models, K-nearest neighbor nonparametric model, and logistic regression model are presented in different sections of this chapter. Each model is supported by related commands and programs provided in *MATLAB*.

Keywords Linear regression • Nonlinear regression • Nonparametric regression • K-nearest neighbor regression • Logistic regression

3.1 Introduction

The aim of regression analysis is to study the relationship between one variable, usually called the dependent variable, and several other variables, called the independent variables often. The independent variables are related to the dependent variable by a parametric or nonparametric function, called the regression function. The regression function involves a set of unknown parameters, which we seek to find in a modeling process. Regression modeling is particularly valuable when the dependent variable is costly or difficult to measure or in some cases impossible to measure. Regression analysis helps one understand how the typical value of the dependent variable changes when any one of the independent variables is varied, while the other independent variables are held fixed.

S. Araghinejad, *Data-Driven Modeling: Using MATLAB® in Water Resources and Environmental Engineering*, Water Science and Technology Library 67,
DOI 10.1007/978-94-007-7506-0_3,

Table 3.1 A summary on the application of regression models in water resources and environmental engineering

Field of the study	Researchers
Urban water management	Adamowski et al. (2012)
Water quality	Abaurrea et al. (2011)
Hydroclimatology	Hu et al. (2011)
Hydroclimatology	Muluye (2011)
Groundwater	Ozdemir (2011)
Storm water management	Tran et al. (2009)
Hydrometry	Petersen-Øverleir (2006)
Groundwater	Cooley and Christensen (2006)
Hydrology	Regonda et al. (2006)
Hydrology	Mehrotra and Sharma (2006)
Climatology	Gangopadhyay et al. (2005)
Water quality	Qian et al. (2005)
Reservoir planning	Adeloye et al. (2003)
Climatology	Sokol (2003)
Climatology	Rajagopalan and Lall (1999)
Hydrology	Demetracopoulo (1994)
Sediment estimation	Crawford (1991)

In the field of water resources and environmental engineering, regression analysis is widely used for prediction, forecasting, estimation of missing and censored data, and, in general, interpolation and extrapolation of data. Regression analysis is also used to understand which among the independent variables are related to the dependent variable and to explore the forms of these relationships. Examples of regression-based models in the field of water resources and environmental engineering are presented in Table 3.1.

In this chapter, Sect. 3.1 discusses linear regression methods, one of the most famous data-driven models and includes point and interval estimation of a dependent variable by independent variables. Section 3.2 discusses nonlinear regression models along with the examples of their application in water resources and environmental engineering. Section 3.3 is concerned with a procedure called nonparametric regression. The famous *K*-nearest neighbor method is presented in this section. Another nonlinear model, called logistic regression, comes in Sect. 3.4. Finally, the chapter ends with a workshop (Fig. 3.1).

3.2 Linear Regression

Linear regression, the most famous data-driven method, is used to model a linear relationship between a continuous dependent variable *Y* and one or more independent variables *X*. Most applications of regression aim to identify what variables are associated with *Y*, to postulate what causes *Y*, to predict future observations of *Y*, or to assess control over a process or system. Generally, explanatory or "causal" models are based on data obtained from well-controlled experiments, quality control models, and observations.

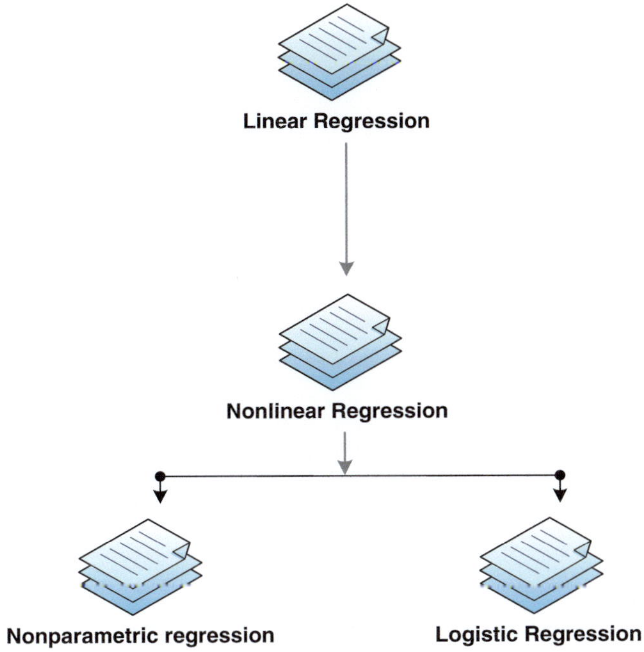

Fig. 3.1 The general structure of contents presented in this chapter

If x and y denote the two variables under consideration, then a scatter diagram shows the location of the points (x,y) as demonstrated in Fig. 3.2. If all points lie near a line, then correlation is called linear (Fig. 3.2a, b). If y tends to increase as x increases, then correlation is called positive (Fig. 3.2a). If y tends to decrease as x increases, then correlation is called negative (Fig. 3.2b). If all points seem to lie near a curve, then correlation is called nonlinear (Fig. 3.2c). The correlation between X and Y is measured by correlation coefficient which is a number in the range of $[-1,1]$ representing different forms of the correlation shown in Fig. 3.2.

The strength of the linear relationship between two variables is measured by the simple correlation coefficient. Correlation coefficient between n observations of X and Y is calculated as

$$R(X,Y) = \frac{\mathrm{Cov}(X,Y)}{\sqrt{\mathrm{Cov}(X,X) \times \mathrm{Cov}(Y,Y)}} \tag{3.1}$$

whcrc

$$\mathrm{Cov}(X,Y) = \sqrt{\frac{1}{n-1}\sum^{n}(x-\overline{x})(y-\overline{y})} \tag{3.2}$$

and n is the number of observations of X and Y values.

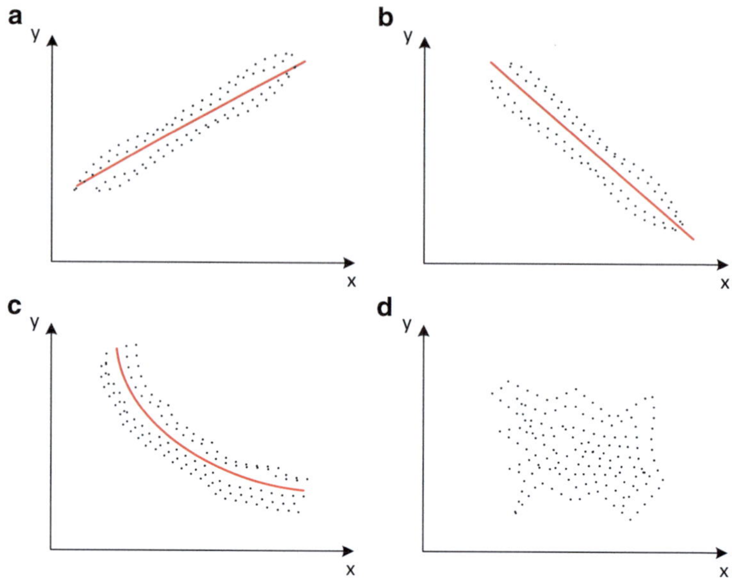

Fig. 3.2 Different types of correlation between Y and X. (**a**) Positive linear correlation (**b**) negative linear correlation (**c**) nonlinear correlation (**d**) no correlation

The correlation coefficient is calculated by the following syntax:

```
R=corrcoef(X)
```

where X is a $n \times m$ matrix; n is the number of observation; and m is the number of variables. R is a $m \times m$ matrix containing the correlation between each pairs of variables. It is to be notified that the small value of the above correlation coefficient does not necessarily mean that two variables are independent. As an example two correlated data of X and Y which are related by function $Y = \sin(2\pi X)$ are shown in Fig. 3.3. The data shown in this figure are completely correlated; however, the correlation coefficient results in 0 for those specific data. It demonstrates that the correlation coefficient only represents linear correlation and might not be much trusted in case of nonlinear relationship between two sets of data.

The simplest regression function is presented as

$$y = \beta_0 + \beta_1 x \tag{3.3}$$

where y is the dependent variable, x is the independent variable, and a and b are the parameters of the model. The above equation seems too simple to be applied in many water resources and environmental cases; however, it is still one of the most popular data-driven models because of two reasons: the well-known and easy methods to calculate the unknown parameters of the equation and the role of this

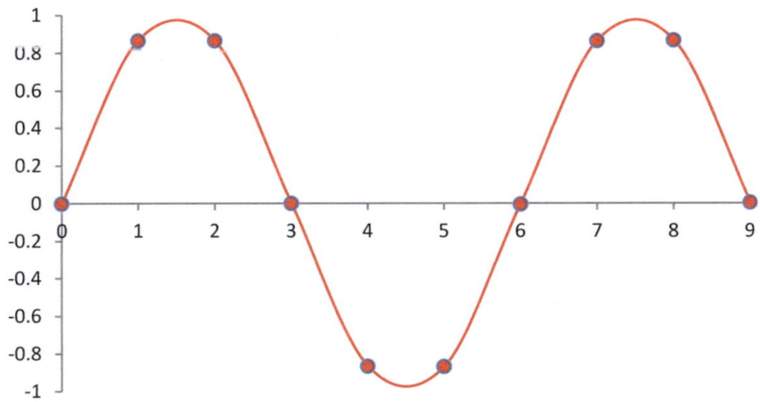

Fig. 3.3 An example of two correlated data with $R = 0$ (the x axis contains and y axis contains)

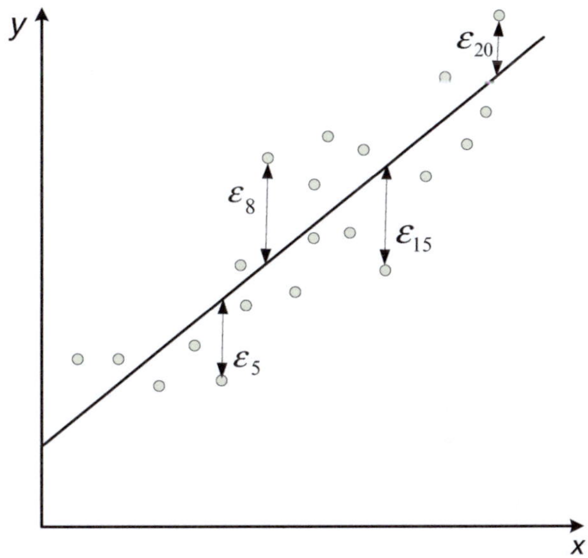

Fig. 3.4 Samples of errors in linear regression fitting

basic model as a baseline for comparing the skill of other models. The process of calculating the parameters of the model shown in Eq. 3.3 is simplified as follows. The aim of linear regression modeling for a set of input/output data is to fit a line in a way that the sum of absolute errors of fitting for n observations (as shown in Fig. 3.4) is minimized.

It means that the following value should be minimized:

$$S^2 = \sum_{i=1}^{n} \varepsilon_i^2 = \sum_{i=1}^{n} (y_i - \beta_0 - \beta_1 x_i)^2 \tag{3.4}$$

In the above equation β_0 and β_1 are the parameters of the model. Solving the following equations gives the best parameters, which minimize S^2:

$$\frac{\partial S^2}{\partial \beta_0} = 0 \quad \text{and} \quad \frac{\partial S^2}{\partial \beta_1} = 0 \tag{3.5}$$

The equations result in

$$-2\sum_{i=1}^{n}\left(y_i - \widehat{\beta}_0 - \widehat{\beta}_1 x_i\right) = 0 \tag{3.6}$$

and

$$-2\sum_{i=1}^{n}\left(y_i - \widehat{\beta}_0 - \widehat{\beta}_1 x_i\right)x_i = 0 \tag{3.7}$$

which gives

$$\beta_1 = \frac{n\sum xy - \sum x \sum y}{n\sum x^2 - \left(\sum x\right)^2} \tag{3.8}$$

and

$$\beta_0 = \overline{y} - \beta_1 \overline{x} \tag{3.9}$$

where $\overline{y}$ and $\overline{x}$ are the average of observed y and x over n observations, respectively.

Multiple linear regression (MLR) is a multivariate method of linear regression which is actually an extended version of Eq. 3.3 presenting as

$$y = \beta_0 + \beta_1 x_1 + \beta_2 x_2 + \cdots + \beta_m x_m \tag{3.10}$$

for a case of m independent variables.

3.2.1 Point and Interval Estimation

The dependent variable in a regression model may be estimated in a deterministic or probabilistic sense. A deterministic estimation is a point estimate of a variable, while a probabilistic estimation specifies a probability distribution function for the dependent variable. The estimation probability is a numerical measure of the degree of certitude about the occurrence of an event, conditional on all information utilized in the estimation process. The probabilistic estimation is also known as interval estimation because it usually offers a range of probable estimated variables.

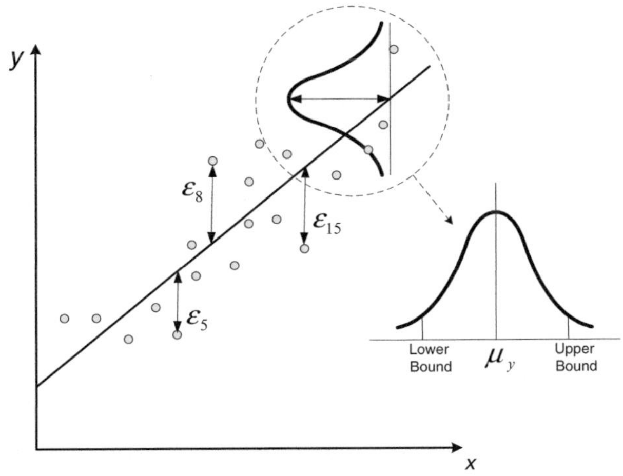

Fig. 3.5 The schematic of a probability distribution function of dependent variable in a linear regression model

In case that the correlation coefficient between two sets of variables is less than one (which almost occurs in all real-world cases related to the fields of water resources and environmental engineering), the estimation of dependent variable involves errors. These estimation errors could be represented in forms of interval estimation instead of the conventional point estimation. In such cases, the deterministic equation of 3.3 is changed to the stochastic form of

$$y = \beta_0 + \beta_1 x + e \tag{3.11}$$

where e is the estimation error.

According to Fig. 3.5, the output of a linear regression model can be presented by a distribution function. The expected value of the estimation is in fact the average value of this distribution which is resulted by Eq. 3.11.

$$\mu_y = \beta_0 + \beta_1 x \tag{3.12}$$

The interval estimation is then calculated as a function of variance of the independent and dependent variables, the covariance between them, and of course the significant level to determine the crisp boundaries of the interval. The interval estimation of y by the use of Eq. 3.11 in case that $x = x_0$ is obtained as

$$(\beta_0 + \beta_1 x_\circ) \pm (t_c)_{\alpha/2} S_e \sqrt{\frac{1}{n} + \frac{n(x_\circ - \overline{x})^2}{S_{xx}}} \tag{3.13}$$

where $(t_c)_{\alpha/2}$ is the critical value at significance level of α and degree of freedom $v = n - 1$ obtained by t-student distribution function and

$$S_{xx} = n\sum x^2 - \left(\sum x\right)^2 \tag{3.14}$$

$$S_{yy} = n\sum y^2 - \left(\sum y\right)^2 \tag{3.15}$$

$$S_{xy} = n\sum xy - \sum x\sum y \tag{3.16}$$

and

$$S_e^2 = \frac{S_{xx}S_{yy} - \left(S_{xy}\right)^2}{n(n-2)S_{xx}} \tag{3.17}$$

Also the confidence limits for regression coefficients can be obtained as

$$\beta_0 \pm (t_c)_{\alpha/2} S_e \sqrt{\frac{S_{xx} + \left(n\overline{X}\right)^2}{nS_{xx}}} \tag{3.18}$$

and

$$\beta_1 \pm (t_c)_{\alpha/2} S_e \sqrt{\frac{n}{S_{xx}}} \tag{3.19}$$

In MATLAB parameters of a linear regression are obtained by the following syntax where *betahat* is the regression parameter:

```
betahat=X\Y;
```

where *X* and *Y* are the set of independent and dependent variables, respectively. In case of interval estimation, the above command changes to the following form:

```
[betahat,Ibeta,res,Ires,stats] = regress(Y,X,alpha);
```

Ibeta is the range of parameters for a multiple linear regression equation. *Res* and *Ires* represent the residuals and the interval of residuals, respectively, and finally *stats* is a vector containing some important statistics including the regression coefficient which appears as the first variable of this vector. To run the command, except *X* and *Y* matrices, we need to introduce *alpha*. *Alpha* is the significant level for calculating the parameters of regression.

Any dependent variable, *Y*, corresponding to the given *X* is calculated by the obtained regression parameters using the following syntax:

```
Yestimate=X*betahat
```

Table 3.2 Data presented for Example 3.1

Distance	TDS	Distance	TDS
10	105	110	233
20	137	120	235
30	188	130	238
40	198	140	244
50	200	150	246
60	202	160	250
70	210	170	253
80	208	180	255
90	218	190	255
100	233	200	260

The intervals of the residuals of the regression fitting (also called the error of the model) could be displayed in MATLAB by the following command:

```
rcoplot(res,Ires);
```

The statistically suspicions outliers are also highlighted in the plot by the "red" color.

Example 3.1: Interpolation of Water Quality Values

Water quality of a river as a function of distance from the upstream of river is tabulated as follows. Use a linear regression model to interpolate *total dissolved solution* (TDS) at different locations of the river in Table 3.2.

Solution
First, we divide the data into two sets of "calibration" and "validation." First 15 data are used for calibration and the rest are used for validation.

```
X1=[ones(size(Xcal,1),1), Xcal];
beta=X1\Ycal;
for i=1:size(Ycal,1); Yhatcal(i,1) = sum(beta'.*[1,
Xcal(i,:)]); end;
RMSEcal=sqrt(sum((Ycal-Yhatcal).^2)/size(Ycal,1))
X2=[ones(size(Xtest,1),1), Xtest];
for i=1:size(Ytest,1); Yhattest(i,1) = sum(beta'.*[1,
Xtest(i,:)]); end;
RMSEtest=sqrt(sum((Ytest-Yhattest).^2)/
size (Ytest,1))
```

Please note that a column of unit variables is added to the input matrix to use the regression form of TDS $= \beta_0 + \beta_1$ Dist, where TDS $=$ total dissolved solids and Dist $=$ distance from the upstream of the river.

The parameters of the equation are obtained as $\beta_0 = 143.59$ *and* $\beta_1 = 0.784$. Root mean square error (RMSE) for calibration data set is obtained as *17.71*, where RMSE for validation data set is *31.21*. RMSE is a common statistic to measure the skill of a regression model. The formulation of calculating this statistic is written in the last sentence of the program of Example 3.1.

Example 3.2: Interval Estimation of Water Quality Values

Solve Example 3.1 in a probabilistic manner and calculate the probable range of TDS at the distance of 125 km from the upstream.

Solution

The following program is used to solve this example considering an additional column of ones for Xcal:

```
alpha=0.05;
[betahat,Ibeta,res,Ires,stats]=...
regress(Ycal,Xcal,alpha);
Yint=[1 125]*Ibeta;
Yestimate=[1 125]*betahat
```

The results would be an interval of [188.6 294.7]. The expected value of the estimation is 241.6.

3.2.2 Preprocessing of Data

Before developing a regression model, the input data might be prepared before being introduced to the model. The following common processes are usually applied for this purpose:

3.2.2.1 Standardizing

Standardizing helps rescaling the input values to a similar scale. The following command standardizes the inputs to fall in the range [−1,1]. This preprocessing task helps modeling process by bringing all inputs to a standard range which makes them comparable easily.

```
[P,PS] = mapminmax(x1)
```

where *PS* contains the process setting and *P* is the transformed input. The transformed matrix could be reversed to the original vector by the following command:

```
x1_again = mapminmax('reverse',P,PS)
```

3.2.2.2 Normalizing

The following command normalizes a vector to have zero mean and unity variance, which is another way to rescale the inputs:

```
[P,PS] = mapstd(x1)
```

The transformed matrix could be reversed to the original vector by the following command:

```
x1_again = mapstd('reverse',P,PS);
```

3.2.2.3 Principal Component Analysis

The principal component analysis (PCA) detects linear dependencies between variables and replaces groups of correlated variables by new uncorrelated variables, called the principal components (PCs).

The *xy* coordinate system can be replaced by a new orthogonal coordinate system, where the first axis passes through the long axis of the data scatter and the new origin is the bivariate mean. This new reference frame has the advantage that the first axis can be used to describe most of the variance, while the second axis contributes only a little. Originally, two axes were needed to describe the data set prior to the transformation. Therefore, it is possible to reduce the data dimension by dropping the second axis without losing much information in case of significantly correlation.

Suppose that we have p vectors of correlated vectors, X_i. Principal components are linear combination of those p vectors using different parameters for each component. For p vectors we will have p principal components; however, the value of each principal component is different from the others in terms of their contribution to the total information of the data (Fig. 3.6).

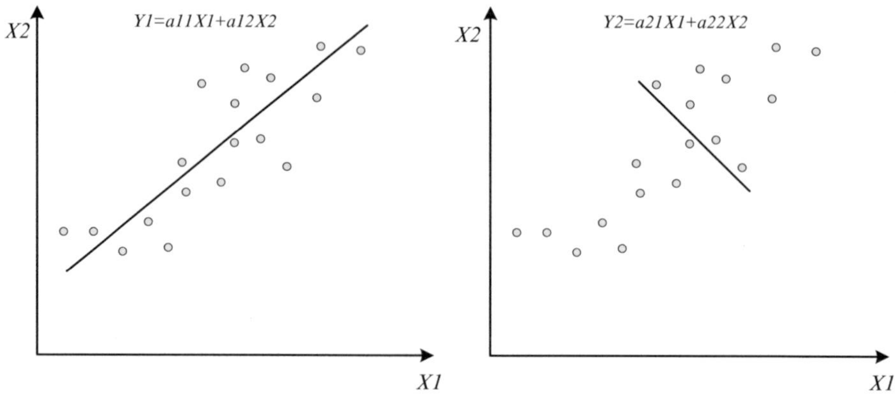

Fig. 3.6 An example of principal component analysis for a two-dimensional data

p principal components are calculated as

$$\begin{aligned}
&PC_1 = Y_1 = a_{11}X_1 + a_{12}X_2 + \ldots a_{1p}X_p \\
&PC_2 = Y_2 = a_{21}X_1 + a_{22}X_2 + \ldots a_{2p}X_p \\
&\vdots \\
&PC_p = Y_p = a_{p1}X_1 + a_{n2}X_2 + \ldots ap_pX_p
\end{aligned} \tag{3.20}$$

The first principal component contains the greatest variance, the second highest variance, and so forth. All *PCs* together contain the full variance of the data set. The variance is concentrated in the first few *PCs*, which explains most of the information content of the data set. The last *PCs* are generally ignored to reduce the data dimension.

The following command calculates principal components of an input matrix. Furthermore, the command enables omitting the *PCs* that contribute to the value of the whole data less than the specific value of "*maxfrac*." "*maxfrac*" is between 0 and 1, representing 0–100 % of the value of the whole data.

It should be notified that it is usual to standardize the data before being used through the process of principal component analysis as follows:

```
[pn,ps1] = mapstd(P)
[ptrans,ps] =processpca(pn,maxfrac);
```

Example 3.3: Principal Components Analysis

Calculate the principal components of the 3-dimentional data shown in Table 3.3.

Table 3.3 Data presented for Example 3.3

P1	P2	P3
40	14	66
45	11	71
70	17	96
30	9	56
23	12.3	49
42	14.2	68
12	10	38
34	13.4	60
56	17	82
98	23	124

Solution

First, to examine the correlation between the three vectors, the matrix of correlation coefficient is calculated using the following command:

```
R—corrcoef(P)
```

where P is the 10 by 3 matrix as presented.

The data results in the following coefficients, which demonstrates a significant correlation between them:

```
R =

    1.0000   0.9124   1.0000
    0.9124   1.0000   0.9124
    1.0000   0.9124   1.0000
```

The data turns into its principal components by the following commands. Note that the data should be transposed before being used through the calculation of PCA.

```
P=P';
[pn,ps1] = mapstd(P)
[ptrans,ps2] = processpca(pn,0.1);
```

The third principal component is omitted because it contains less than 10 % of the information within the entire data.

Table 3.4 Conventional nonlinear regression functions

Model	Equation
Exponential	$y = ae^{bx} + ce^{dx}$
Fourier	$y = a_0 + a_1 \cos(x * w) + b_1 \sin(x * w) + \cdots a_n \cos(x * w) + b_n \sin(x * w)$
Gaussian	$y = a_1 e^{[(x-b_1)/c_1]^2} + \cdots + a_n e^{[(x-b_n)/c_n]^2}$
Polynomial	$y = a_1 x^n + a_2 x^{n-1} + \cdots + a_n x + a_0$
Power	$y = ax^b + c$
Sin function	$y = a_1 \sin(b_1 x + c_1) + \cdots + a_n \sin(b_n x + c_n)$
Weibull	$y = abx^{b-1} e^{(-a*x^b)}$

The correlation coefficient between two remaining principal components (the first and the second one) would be as follows which demonstrates that they are independent:

```
R =

    1.0000   0.0000
    0.0000   1.0000
```

3.3 Nonlinear Regression

Nonlinear regression is a term that is used for a wide range of regression models which all present a nonlinear relationship between input and output data. Different mathematical functions are applied in the field of nonlinear regression, ranging from sinusoidal function to exponential and power ones. Table 3.4 shows a list of regression functions that could be used as nonlinear regression models in MATLAB. x and y represent independent and dependent variables, respectively. The others are the parameters of the model.

Two commands could be used to make the application of nonlinear regression possible:

```
cftool
```

presenting a tool for curve fitting and

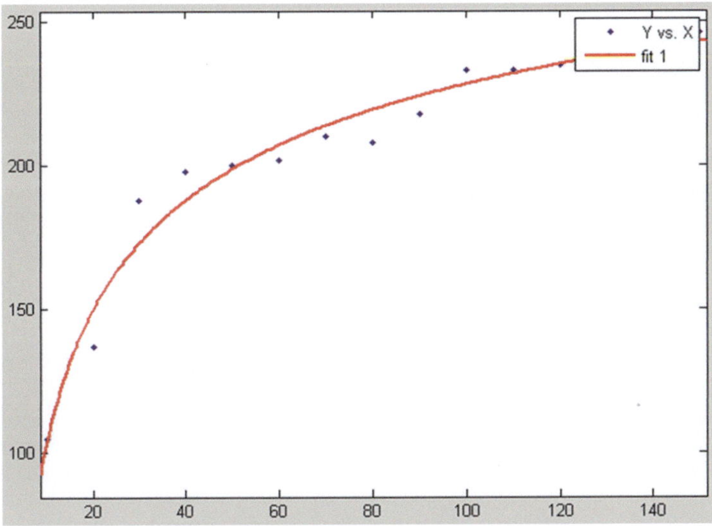

Fig. 3.7 The power equation fitted to the data of Example 3.4

```
sftool
```

presenting a tool for surface fitting. An example of nonlinear regression fitting is presented in the following example.

Example 3.4: Nonlinear Regression Modeling

Fit a nonlinear equation to solve the problem of Example 3.1.

Solution
We chose a power equation to fit the data as follows:

$$y = f(x) = ax^b + c$$

Fitting the equation on the calibration data set (first 15 data) results in the following parameters:

$a = -498.9$
$b = -0.2821$
$c = 364.3$

The equation is fitted on the data with a high correlation coefficient of $R = 0.9829$ (Fig. 3.7).

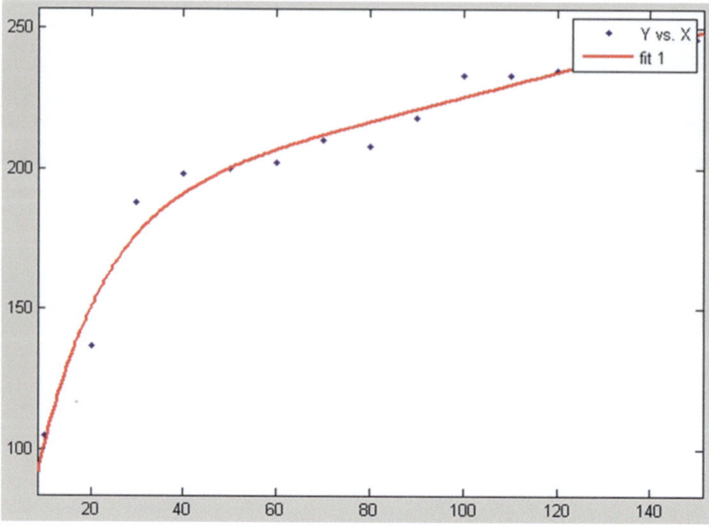

Fig. 3.8 The exponential equation fitted to the data of Example 3.4

Changing the equation to an exponential form of $y = f(x) = ae^{bx} + ce^{dx}$ with the fitted parameters of

$a = 186.6$
$b = 0.001894$
$c = -182$
$d = -0.07209$

will result in the correlation coefficient of 0.9836 which is a little bit higher than what is obtained by the previous equation. Figure 3.8 shows the fitted equation on the calibration data.

The skill of both models could be tested on the validation data set.

In addition to use the curve and surface fitting tools, specific nonlinear functions could be applied to pairs of input/output data by the use of the command line. For example, polynomial function of order n can be fitted to the data by the following syntax:

```
[p] =polyfit(X,Y,n);
```

where n is the order of the model and p contains the parameters of $y = a_1x^n + a_2x^{n-1} + \cdots + a_nx + a_0$.

Table 3.5 Estimated variables and errors in Example 3.5

X	*Y*	Estimated *Y*	Error
10.0	105.0	135.4	−30.4
20.0	137.0	149.5	−12.5
30.0	188.0	162.7	25.3
40.0	198.0	175.0	23.0
50.0	200.0	186.4	13.6
60.0	202.0	196.9	5.1
70.0	210.0	206.6	3.4
80.0	208.0	215.3	−7.3
90.0	218.0	223.1	−5.1
100.0	233.0	230.1	2.9
110.0	233.0	236.1	−3.1
120.0	235.0	241.3	−6.3
130.0	238.0	245.6	−7.6
140.0	244.0	249.0	−5.0
150.0	246.0	251.5	−5.5
160.0	250.0	253.1	−3.1
170.0	253.0	253.8	−0.8
180.0	255.0	253.6	1.4
190.0	255.0	252.5	2.5
200.0	260.0	250.5	9.5

Example 3.5: Polynomial Regression

Fit a polynomial equation of order 2 to the data of Example 3.1.

Solution

The following short program is used to fit a quadratic equation to the data. To get a summary report, the fitting errors are tabulated at the end of the program (Table 3.5).

```
n=2;
[p] = polyfit(X,Y,n);
Yhat = polyval(p,X);
table = [X Y Yhat Y-Yhat]
```

3.4 Nonparametric Regression

The recognition of the nonlinearity of the underlying dynamics of stochastic processes, gains in computational capability, and the availability of large data sets have spurred the growth of nonparametric methods. Nonparametric estimation of probability densities and regression functions is pursued through weighted local averages of the dependent variable. This is the foundation for nearest neighbor methods. *K*-nearest neighbor (*K*-NN) methods use the similarity (neighborhood) between observations of predictors and similar sets of historical observations (successors) to obtain the best estimate for a dependent variable (Karlsson and Yakowitz 1987; Lall and Sharma 1996).

Nonparametric regression is a form of regression analysis in which the predictor does not take a predetermined form but is constructed according to information derived from the data. Nonparametric regression requires larger sample sizes than regression based on parametric models because the data must supply the model structure as well as the model estimates. The *K*-NN method imposes a metric on the predictors to find the set of *K* past nearest neighbors for the current condition in which the nearest neighbors have the lowest distance. The distance between the current and historical condition could be calculated by the Euclidian (Karlsson and Yakowitz 1987) or *Mahalanobis* distance (Yates et al. 2003) between current and historical predictors.

The algorithmic procedure of a *K*-NN regression is summarized in Fig. 3.9 and is presented as follows:

- Determine the vector of current m independent variables also known as predictors, $X_r = \{x_{1r}, x_{2r}, x_{3r} \ldots x_{mr}\}$, associated with the dependent variable, Y_r.
- Determine the matrix of $n \times m$ predictors containing n vectors of already observed predictors, $X_t = \{x_{1t}, x_{2t}, x_{3t} \ldots x_{mt}\}$; $t = 1, 2, \ldots, n$.
- Calculate n distances between current predictor and the observed predictors, Δ_{rt}.
- Select K sets of predictors/dependent variables (X_k, Y_k), which have the lowest values of Δ_{rt}. Those sets are known as the K-nearest neighbors.
- Calculate a kernel function associated with each K-nearest neighbor as

$$f_k(\Delta_{rk}) = \frac{1/\Delta_{rk}}{\sum_{k=1}^{K} 1/\Delta_{rk}} \tag{3.21}$$

Obviously, $\sum_{k=1}^{K} f_k(\Delta_{rk}) = 1$.

- The unknown Y_r is finally calculated as

$$Y_r = \sum_{j=1}^{K} f_k(\Delta_{rk}) \times Y_k \tag{3.22}$$

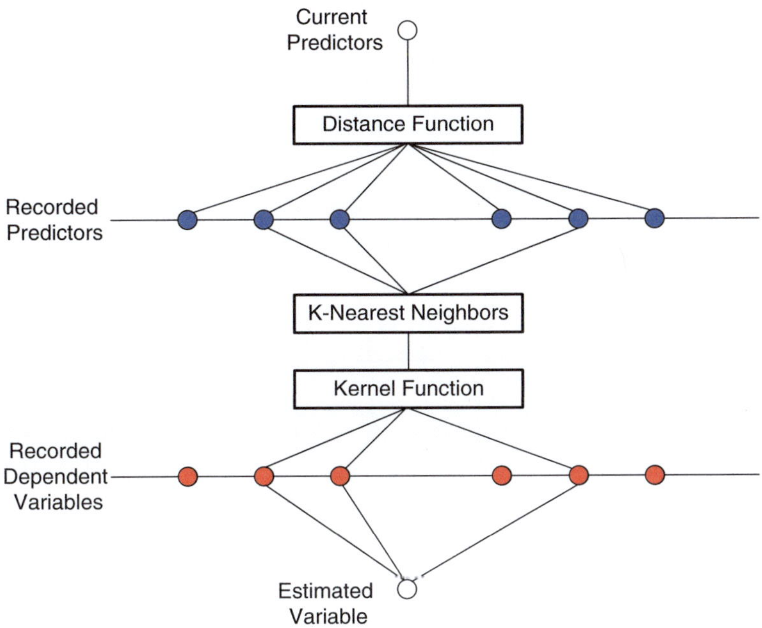

Fig. 3.9 The schematic of *K*-NN algorithm

The distance function is usually calculated by a *Euclidean* distance or a *Mahalanobis* distance. A Euclidean distance between *i*th and *j*th predictors is calculated as

$$\Delta_{ij} = \sqrt{\left(x_{1i} - x_{1j}\right)^2 + \left(x_{2i} - x_{2j}\right)^2 + \cdots + \left(x_{mi} - x_{mj}\right)^2} \tag{3.23}$$

where m is the dimension of the predictors. The *Mahalanobis* distance uses the following equation:

$$\Delta_{ij} = \sqrt{\left(X_i - X_j\right)C_t^{-1}\left(X_i - X_j\right)^T} \tag{3.24}$$

where C is the covariance matrix between X and Y.

Mahalanobis distance is a distance measure introduced by *Mahalanobis* in 1936 (Mahalanobis 1936). It is based on correlations between variables by which different patterns can be identified and analyzed. It differs from Euclidean distance in that it takes into account the correlations of the data set and is scale invariant.

Lall and Sharma (1996) suggested that instead of the kernel function of

$$f_k(\Delta_{rk}) = \frac{1/\Delta_{rk}}{\sum_{k=1}^{K} 1/\Delta_{rk}},$$

Table 3.6 Data presented for Example 3.6

X	y
2	4
3	9
4	16
5	25
7	49
8	64
6	?

Table 3.7 Solution table for Example 3.6

1	2	3	4	5	
$6-2=4$	1	49	1.000	0.333	16.33
$6-3=3$	1	25	1.000	0.333	8.33
$6-4=2$	2	64	0.500	0.167	10.66
$6-5=1$	2	16	0.500	0.167	2.66
$7-6=1$	3	9			Sum = 38
$8-6=2$	4	4			

the following function could be used:

$$f_k(j) = \frac{1/j}{\sum_{j=1}^{K} 1/j} \tag{3.25}$$

where j is the order of the neighbors after sorting them in an ascending order. Neighbors with higher distance get higher orders and the lower contribution to the final output.

Example 3.6: An Illustrative Example of Using K-NN Regression

Table 3.6 shows dependent variable y which is related to the input x by the simple equation of $y = x^2$. Use the K-NN method to estimate the output associated to $x = 6$.

Solution

Table 3.7 shows the procedure for applying K-NN to solve the problem. Column 1 demonstrates the distance between the current predictor, $x = 6$, and the observed predictors. Column 2 sorts the recorded data according to the distance of their predictors to the current predictor. Column 3 shows the dependent variables associated to the neighbors. Column 4 demonstrates the inverse of distances. Four nearest records are considered as the neighbors. Column 5 demonstrates

Table 3.8 Data presented for Example 3.7

X	Y
2	4
3	9
4	16
5	25
7	49
8	64
6	?

the kernel functions. And finally column 6 shows the contribution of each recorded data to the unknown variable. The final results are in fact a summation of these contributions.

It should be notified that the result of 38 is associated to 4 number of neighbors ($K = 4$). Changing this parameter might change the final result.

The advantages of using a *K*-NN regression are:

- It follows a simple algorithm.
- It does not need any specific assumption.
- It is used for both linear and nonlinear problems.

The *K*-NN has some limitations including:

- It is basically an interpolation algorithm which might result in significant errors in case of extreme value estimation.
- *K-NN* is considered as a nonparametric approach. It means that *K*-NN does not use any parametric distribution. However, there are still some parameters that might affect the skill of a *K*-NN model. Most important parameters are the weights *Euclidean* distance and, most importantly, the number of contributing neighbors, *K*.

The following examples present an approach for determining the appropriate parameters for a specific problem:

Example 3.7: Development of a K-NN Regression Model

Develop a *K*-NN regression model and find the dependent variable associated to $x = 6$ using the data of Table 3.8.

Solution

The following program finds the best weights for the predictors and the best number of neighbors for a certain data to be used in *K*-NN method. Then, it calculates the results of *K*-NN regression by those parameters. The following comments should be considered before using the program:

- The program has been developed for 1- and 2-dimentional inputs and 1-dimentional dependent variable.
- Input and dependent variables are introduced by *X* and *Y* matrices, respectively.
- The program asks the number of observations of *X* and *Y* and calculates the best weights (Best_W) and the best number of neighbors (Best_K) in a way to calculate the minimum estimation error for *X* and *Y* matrices.
- The program calculates the estimated dependent variable associated to a specific "current state of predictors," which should be introduced by a certain matrix by the user.

It should be notified that at each calculation only one observation is introduced as the current state of predictors. The program does not work for a current predictor which is similar to the ones in the *X* matrix, because the similar inputs are supposed to give the results equal to what has already been observed in *Y* matrix.

```
% K-NN regression for 1- and 2- dimensional input vari-
ables and
% 1-dimensional dependent variable

X=X; % Input Variables
Y=Y; % Dependent Variables

%% Calculating the best "K" in K-NN regression
%The inputs to the Program are X and Y matrices
Nom=input('Enter number of time steps=');
f=0;
for n=1:Nom
  Xtest=X(n,:);
  Ytest=Y(n);
  for m =1:1:Nom-1
    if m<n
      B(m,:)= X(m,:);
      YB(m,1)= Y(m);
    else
      B(m,:)= X(m+1,:);
      YB(m,1)= Y(m+1);
    end
  end

  % The calculation starts with k=2
```

(continued)

```
  for k=2:Nom-1
    for W_1=0:0.1:1.0
      W(1)= W_1;
      if size(X,2)== 1
        W(1)= 1;
        f=4*(n-1)+ k-1;
      else
        W(2)= 1-W(1);
        f=f+1;
      end
      d=zeros(n-1,1);
      result=0;
      finalresults=0;
      for j=1:Nom-1
        d(j,1)= sqrt(sum((W.*(B(j,:)-Xtest).^2)));
      end
      [sortedd firstindex]=sort(d,1,'ascend');
      sumd=sum(1./sortedd(1:k));
      prob=(1./sortedd(1:k))./sumd;
      result(1:k)=YB(firstindex(1:k)).*prob(1:k);
      finalresults(k)=sum(result);
         error(k,f)=abs((finalresults(k)-Ytest)/Ytest)
*100;
      Table(f,1)=k;
      Table(f,2:size(X,2)+1)=W(:);
      Table(f,size(X,2)+2)=error(k,f);
    end
  end
end
Table2=sortrows(Table);

w=0;
for k=2:Nom-1
  for W_1=0:0.1:1.0
    W(1)=W_1;

    if size(X,2)==2
      W(2)=1-W(1);
      w=w+1;
    else
      W(1)=1;
      w=k-1;
    end
```

(continued)

```
    meanerror(w+(k-2)*(11*size(X,2)-11),1)=k;
meanerror(w+(k-2)*(11*size(X,2)-11),2:size(X,2)+1)=
W(1:size(X,2));
       meanerror(w+(k-2)*(11*size(X,2)-11),size(X,2)
+2)=...
      mean(Table2(1+(k-2)*(Nom*(11*size(X,2)-11))+...
(w-1)*Nom:(k-2)*(Nom*(11*size(X,2)-11))+(w)*Nom,
size(X,2)+2));
  end
  w=0;
end
leasterror= min(meanerror(:,size(X,2)+2));

g=0;
for g=1:(Nom-2)*((11*size(X,2)-11)-(size(X,2)-2))
  if meanerror(g,size(X,2)+2)==leasterror
    Final(1,size(X,2)+2)=fprintf('Least Error');
    Final(1,1)=meanerror(g,1);
      Final(1,2:size(X,2)+1)=meanerror(g,2:size(X,2)
+1);
    Final(1,size(X,2)+2)=meanerror(g,size(X,2)+2);
    Best_K = Final(1,1)
    if size(X,2)==2
    Best_W=Final(1,2:size(X,2)+1)
    end
    LeastError=Final(1,size(X,2)+2)
  end
end

%% KNN calculation%%
k=Best_K;
if size(X,2)==2
  W=Best_W;
else
  W=1;
end

current=input('Enter the current state of predictors=');
Nom=input('Enter number of time steps=');
%% Calculating drh%%
d=zeros(Nom,1);
```

(continued)

```
for j=1:Nom
  d(j)=sqrt(sum((W.*(X(j,:)-current).^2)));
end

%% Sorting drh in an ascending order%%
[sortedd firstindex]=sort(d,1,'ascend');
sumd=sum(1./sortedd(1:k));
prob=(1./sortedd(1:k))./sumd;
%% Calculating Yr%%
result=zeros(k,1);
result(1:k)=Y(firstindex(1:k)).*prob(1:k);
finalresult=sum(result)
```

To use the above program, the following matrices need to be set:

X = input data
Y = recorded dependent data
P = current state of predictors = 6
Number of time steps = 6

Then the results will be:

Best_K = 2 = best number of neighbors
Final result = dependent variable associated to the current predictor = 37

3.5 Logistic Regression

Logistic regression is a class of regression models which deals with the proportion data. In case that the dependent variable is a probability of an event (a number between 0 and 1), use of a linear regression function may lead to inappropriate results mostly beyond the acceptable range of [0,1]. Suppose that we need to set up a regression model to estimate the flood condition based on the observed and forecast rainfall. The occurrence of a flood event is in fact a binary variable that could occur when the rainfall exceeds a threshold. The following diagram shows an illustrative example for this case. If we consider p as the probability of flood occurrence, the relationship between p and a linear function in form of $p = f(x)$ is not a good choice at all to fit the data as demonstrated in Fig. 3.10.

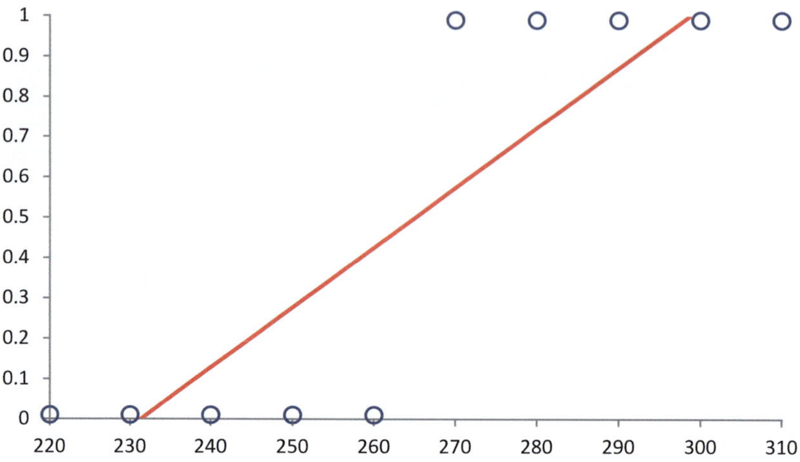

Fig. 3.10 An example of fitting a linear function to a binary data which does not result in appropriate estimation (the x axis contains rainfall values and the y axis contains probability of flooding)

To solve the problem, the linear regression function of $p = f(x)$ is changed to the linear regression function of $\log\left(\frac{p}{1-p}\right) = f(x)$ or

$$\log\left(\frac{p}{1-p}\right) = b_1 + b_2 x \tag{3.26}$$

Equation 3.26 is called a logistic regression which outperforms conventional linear regressions in two ways:

- It better fits on proportion or binary data.
- It avoids huge errors and out of range variables in extreme value estimation.

Example 3.8: Forecasting the Probability of Flooding

The following table shows the data related to the long-term records of rainfall depth and flood events in a region. The first column shows the different values of rainfall depth; meanwhile, the second column presents the number of observation of that rainfall in the historical records of the region. Finally, the third column shows the number of flood events that have occurred associated to the rainfall events. Develop a logistic regression to estimate the probability of flood occurrence associated to the rainfall forecast (Table 3.9).

Solution
First, we plot rainfall versus the probability of flood occurrence (Fig. 3.11).

Table 3.9 Data presented for Example 3.8

Depth of rainfall	Number of observation	Number of flood events that occurred
210	48	1
230	42	2
250	31	0
279	34	3
290	31	8
310	21	8
330	23	14
350	23	17
370	21	19
390	16	15
410	17	17
430	21	21

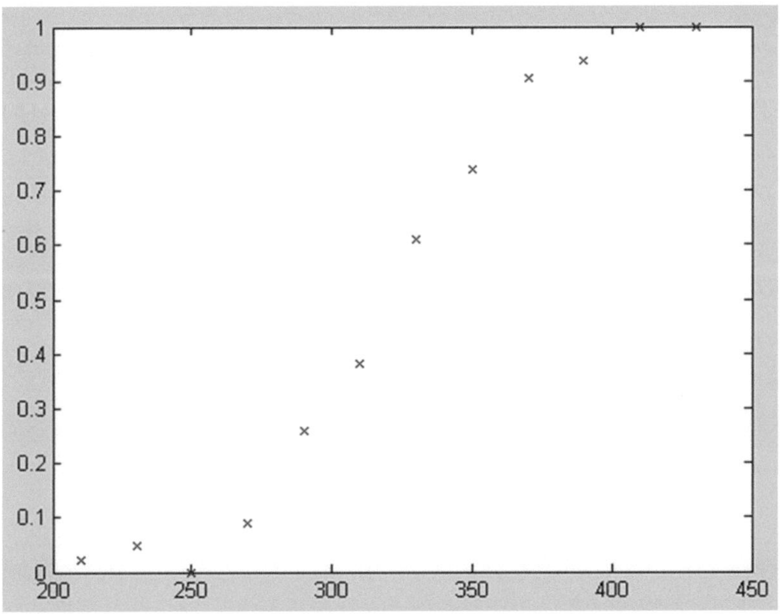

Fig. 3.11 Rainfall (x axis) versus the probability of flood occurrence (y axis)

```
rainfall = [210 230 250 270 290 310 330 350 370 390 410
430]';

flood = [1 2 0 3 8 8 14 17 19 15 17 21]';

observation = [48 42 31 34 31 21 23 23 21 16 17 21]';

plot(rainfall,flood./observation,'x')
```

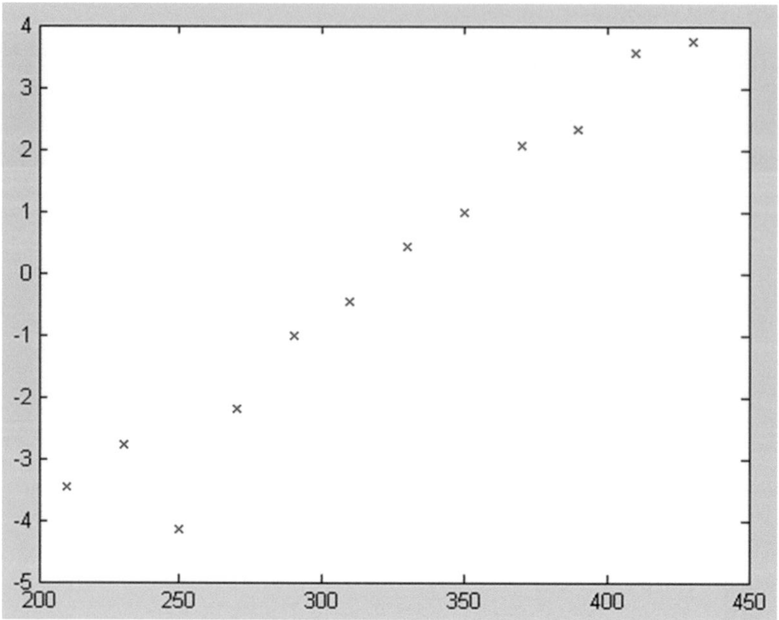

Fig. 3.12 Rainfall (x axis) versus $\log\left(\frac{p}{1-p}\right)$ (y axis), where p is the probability of flooding

Now rainfall versus $\log\left(\frac{p}{1-p}\right)$ is plotted by the following commands (Fig. 3.12):

```
padj = (flood+.5) ./ (observation+1);
plot(rainfall,log(padj./(1-padj)),'x')
```

A logistic regression is fitted to the data by the following syntax:

```
b = glmfit(rainfall,[flood observation],'binomial')
```

which results in the following parameters of the model and the final equation:

```
b =

   -13.3801
     0.0418
```

and $\log\left(\frac{p}{1-p}\right) = -13.3801 + 0.0418x$

Table 3.10 Results obtained by solution of Example 3.8

Probability of flood occurrence	Estimated flood probability
0.02	0.01
0.05	0.02
0.00	0.05
0.09	0.11
0.26	0.22
0.38	0.40
0.61	0.60
0.74	0.78
0.90	0.89
0.94	0.95
1.00	0.98
1.00	0.99

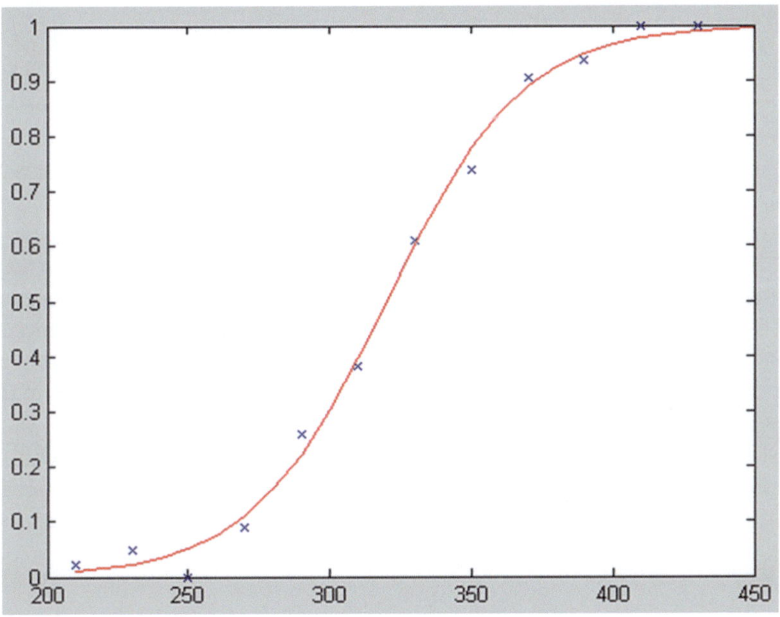

Fig. 3.13 Logistic regression curve fitted to data of Example 3.8

Using "glmval," any output associated with any specific input is obtained by the derived regression model. For instance, considering the same inputs of this example, the estimated outputs are calculated and plotted as shown in Table 3.10 and Fig. 3.13:

```
x= 210:10:450;
y=glmval(b,x,'logit');
plot(rainfall,flood./observation,'x',x,y,'r-')
```

Table 3.11 A review on the highlights and applicable areas of regression models presented in Chapter 3

Model	Highlights	Applicable area
Multiple linear regression	Uses a very efficient structure	Linear problems
Nonlinear regression	Better fits to a data with specific well-known parametric nonlinear curve	Nonlinear problems
K-nearest neighborhood	Uses a simple but very efficient structure Is not suitable for extrapolation	Linear and nonlinear problems
Logistic regression	Transforms a linear equation to an exponential curve to fit a set of S-shape data	Binary or proportional problems

3.6 Summary

Four types of regression models, which are frequently used in the field of water resources and environmental engineering, have been reviewed in this chapter, namely, linear regression, nonlinear regression, nonparametric *K*-nearest neighborhood regression, and logistic regression models. As it is understood by their names, linear and nonlinear regression models are suitable for approximately linear and nonlinear systems, respectively. The *K*-NN regression model can be used either in a linear or nonlinear system. However, it is conventionally used in nonlinear problems. Logistic regression is a type of regression model employed in case of applying proportion or binary data. The estimated variable in a regression model could be produced either by point estimation or by interval estimation. The approach of interval estimation by multiple linear regressions has been reviewed in the chapter. Highlights in the characteristics of the regression models and their area of application are reviewed in Table 3.11.

It should be notified that regression-based models are comparable with those soft-computing models such as artificial neural networks, support vector machines, and fuzzy inference systems that are going to be discussed in the next chapters.

Workshop

Table 3.12 shows the recharge and discharge of an aquifer as well as its water table changes. The volume of the aquifer is a function of recharge and discharge components. We need to develop a general water balance model for the entire aquifer to predict the expected changes of the storage volume. Data of 11 years are considered for the model calibration and 3 years' data are considered for model validation as shown in Tables 3.12 and 3.13.

Solution

- *Fitting a Linear Regression*

Table 3.12 The workshop data used for calibration

No.	Recharge (*R*)	Discharge (*D*)	Change of the water table level (*WT*)
1	433.5	27.0	0.04
2	273.1	31.4	0.00
3	201.8	33.8	1.08
4	235.5	38.0	4.47
5	281.8	33.8	3.16
6	275.0	35.0	3.19
7	277.9	33.8	2.98
8	324.6	37.0	3.15
9	284.0	33.8	2.97
10	224.7	38.0	5.31
11	113.5	39.0	7.42

Table 3.13 The workshop data used for validation

No.	Recharge (*R*)	Discharge (*D*)	Change of the water table level (*WT*)
1	375.82	29.32	1.56
2	163.42	33.82	2.13
3	124.16	35.12	6.01

The following syntax is used to apply a liner regression model to the calibration data set:

```
[betahat,Ibeta,res,Ires,stats] = regress(Y,X,0.05);
```

Please note that to get better results, it is common to add a column of ones to the input data for using the equation of $WT = \beta_0 + \beta_1 \times R + \beta_2 \times D$ where *WT*, *R*, and *D* are water table level, recharge, and discharge, respectively.

Using the input data of

```
X =
  1.0000 433.5000  27.0000
  1.0000 273.1000  31.4000
  1.0000 201.8000  33.8000
  1.0000 235.5000  38.0000
  1.0000 281.8000  33.8000
  1.0000 275.0000  35.0000
  1.0000 277.9000  33.8000
  1.0000 324.6000  37.0000
  1.0000 284.0000  33.8000
  1.0000 224.7000  38.0000
  1.0000 113.5000  39.0000
```

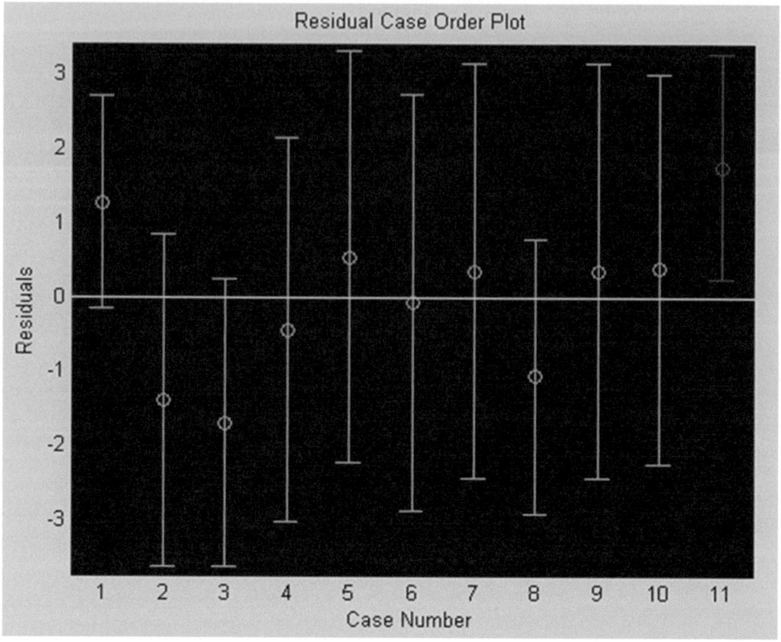

Fig. 3.14 Residual plot for the linear regression fitted to the calibration data set of workshop

results in the following parameters:

```
betahat =

 -14.5232
 -0.0020
  0.5235
```

Using the syntax of

```
rcoplot(res,Ires)
```

the following informative plot of Fig. 3.14 is reflects the estimation errors. The estimated values could be obtained by the following syntax:

```
Yestimate=X*betahat
```

which results in

```
Yestimate =

 -1.2356
  1.3809
  2.7765
  4.9093
  2.6203
  3.2617
  2.6279
  4.2118
  2.6160
  4.9304
  5.6710
```

The validation results are also obtained by the above syntax using the data of Table 3.13.

```
Xval =

  1.0000 375.8200  29.3200
  1.0000 163.4200  33.8200
  1.0000 124.1600  35.1200
```

which results in

```
Yval =

  0.0915
  2.8619
  3.6191
```

The correlation coefficient of the validation data using the linear regression is about 0.75, where the absolute error of estimation, which is obtained as the average of the absolute value of "(actual dependent variable – estimated dependent variable)/actual dependent variable" in percent, is 56.1.

- *Fitting a Nonlinear Regression*

  Using the following syntax

```
sftool
```

two nonlinear equations of $WT = \beta_0 + \beta_1 \times \sqrt{R} + \beta_2 \times D^3$ are fitted to the data which results in the following parameters:

$$WT = -1.771 - 0.1033 \times \sqrt{R} + 0.0001531 \times D^3$$

The correlation coefficient between the actual dependent variable and the estimated variable is 0.902.

The results for three validation data is obtained as

```
0.08535
2.830832
3.709869
```

The correlation coefficient of the validation data using the nonlinear regression is about 0.77, where the absolute error of estimation is 55.2.

- *Fitting a K-NN Regression*

The best number of neighbors is obtained as $K = 10$. The weight for the first variable is obtained as 1 and the weight for the second variable is obtained as 0.

The results for the validation data is obtained as

```
2.44
3.577
5.506
```

The correlation coefficient of the validation data using the K-NN regression is about 0.97, where the absolute error of estimation is 44.2.

References

Abaurrea J, Asin J, Cebrian AC, Garcia-Vera MA (2011) Trend analysis of water quality series based on regression models with correlated errors. J Hydrol 400:341–352

Adamowski J, Chan HF, Prasher SO, Ozga-Zielinski B, Sliusarieva A (2012) Comparison of multiple linear and nonlinear regression, autoregressive integrated moving average, artificial neural network, and wavelet artificial neural network methods for urban water demand forecasting in Montreal, Canada. Water Resour Res 48:W01528 (14)

Adeloye AJ, Lallemand F, McMahon TA (2003) Regression models for within-year capacity adjustment in reservoir planning. Hydrol Sci J 48(4):539–552

Cooley RL, Christensen S (2006) Bias and uncertainty in regression-calibrated models of groundwater flow in heterogeneous media. Adv Water Res 29:639–656

Crawford CG (1991) Estimation of suspended-sediment rating curves and mean suspended-sediment loads. J Hydrol 129:331–348

Demetracopoulo AC (1994) Nonlinear regression applied to hydrologic data. J Irrig Drain Eng 120 (3):652–659

Gangopadhyay S, Clark M, Rajagopalan B (2005) Statistical downscaling using K-nearest neighbors. Water Resour Res 41:W02024 (23)

Hu J, Liu J, Liu Y, Gao C (2011) EMD-KNN model for annual average rainfall forecasting. J Hydrol Eng. doi:10.1061/(ASCE)HE.1943-5584.0000481

Karlsson M, Yakowitz S (1987) Nearest-neighbor methods for nonparametric rainfall-runoff forecasting. Water Resour Res 23(7):1300–1308

Lall U, Sharma A (1996) A nearest neighbor bootstrap for resampling hydrologic time series. Water Resour Res 32(3):679–694

Mahalanobis PC (1936) On the generalised distance in statistics. Proc Natl Inst Sci India 2(1):49–55

Mehrotra R, Sharma A (2006) Conditional resampling of hydrologic time series using multiple predictor variables: a K-nearest neighbour approach. Adv Water Res 29:987–999

Muluye GY (2011) Implications of medium-range numerical weather model output in hydrologic applications: assessment of skill and economic value. J Hydrol 400:448–464

Ozdemir A (2011) Using a binary logistic regression method and GIS for evaluating and mapping the groundwater spring potential in the Sultan Mountains (Aksehir, Turkey). J Hydrol 405:123–136

Petersen-Øverleir A (2006) Modelling stage-discharge relationships affected by hysteresis using the Jones formula and nonlinear regression. Hydrol Sci J 51(3):365–388

Qian SS, Reckhow KH, Zhai J, McMahon G (2005) Nonlinear regression modeling of nutrient loads in streams: a Bayesian approach. Water Resour Res 41:W07012 (10)

Rajagopalan B, Lall U (1999) A k–nearest-neighbor simulator for daily precipitation and other weather variables. Water Resour Res 39(10):3089–3101

Regonda SK, Rajagopalan B, Clark M (2006) A new method to produce categorical streamflow forecasts. Water Resour Res 42:W09501 (6)

Sokol Z (2003) Utilization of regression models for rainfall estimates using radar-derived rainfall data and rain gauge data. J Hydrol 278:144–152

Tran HD, Perera JC, Ng AWM (2009) Predicting structural deterioration condition of individual storm-water pipes using probabilistic neural networks and multiple logistic regression models. J Water Resour Plann Manage 135(6):553–557

Yates D, Gangopadhyay S, Rajagopalan B, Strzepek K (2003) A technique for generating regional climate scenarios using a nearest-neighbor algorithm. Water Resour Res 39(7):1114–1121

Chapter 4
Time Series Modeling

Abstract Modeling of time series involves dealing with the important temporal dimension, which represents and processes sequential inputs. Many statistical-based methods are used to model and forecast time series data such as autoregressive (AR) and autoregressive moving average (ARMA) models, autoregressive integrated moving average (ARIMA) model, and autoregressive moving average with exogenous (ARMAX) data. Time series modeling involves techniques that relate time series data as dependent variables to the predictors, which all are a function of time. Many examples of time series data exist in the field of water resources and environmental engineering, including streamflow data, rainfall data, and time series of total dissolved solids in a river. This variety makes the application of time series very interesting in those fields. Two major applications are usually followed up by the time series modeling: forecasting and synthetic data generation. This chapter reviews the basic mathematical representation as well as the applicable fields of the well-known time series models. In addition to the time series analysis, different models and applications are presented by different programs developed in MATLAB.

Keywords Time series analysis • Time series modeling • AR models • ARMA • ARMAX • Multivariate models

4.1 Introduction

A time series is a sequence of data, which are ordered in time. If observations are made on some phenomenon throughout time, it is most sensible to display the data in the order in which they arose, particularly since successive observations will probably be dependent. It is usual to plot the series value *X* on the vertical axis and time *t* on the horizontal axis to present a time series. Time plays the role of independent variable for the dependent variables of time series data. Such data might be continuous, where we have an observation at every instant of time, e.g., air

S. Araghinejad, *Data-Driven Modeling: Using MATLAB® in Water Resources and Environmental Engineering*, Water Science and Technology Library 67,
DOI 10.1007/978-94-007-7506-0_4,

Table 4.1 A summary review on the application of time series modeling in water resources and environmental engineering

Field of the study	Researchers	Summary
Water quality	Abaurrea et al. (2011)	Trend analysis of water quality time series
Hydroclimatology	Şen (2011)	Trend analysis of data
Hydrology	Shao and Li (2011)	Trend analysis for seasonal time series
Surface reservoir	Song et al. (2011)	Trend analysis of inflow to the reservoir
Hydrology	Cong et al. (2009)	Trend analysis
Hydrology	Hamed (2009a, b)	Trend analysis
Irrigation and drainage	Landeras et al. (2009)	Forecasting evapotranspiration data
Hydroclimatology	Bayazit and Önöz (2007)	Trend analysis
Hydrology	Mohammadi et al. (2006)	Parameter estimation of an ARMA model
Hydrology	Yue and Pilon (2004)	Time series tests
Water quality	Darken et al. (2002)	Serial correlation analysis
Hydrodynamics and water quality	Sun and Koch (2001)	Analysis and forecasting of salinity data
Hydroclimatology	Burlando et al. (1993)	Forecasting of short-term rainfall

temperature, river discharge, and groundwater level. They could be discrete, where we have an observation at some specific times, e.g., flood and drought events.

Time series data are assigned by the time of occurrence, $X = \{X_1, X_2, \ldots, X_t\}$. There are two characteristics that change a vector of data into a time series:

1. The importance of temporal order of data that prevents any changes in the index of the variables
2. Correlation between data of a time series that enables development of mathematical models over the data

Time series modeling involves techniques that relate time series data as dependent variables to the predictors, which all are a function of time. Many examples of time series data exist in the field of water resources and environmental engineering, including streamflow data, rainfall data, and time series of total dissolved solids in a river. It makes the application of time series very interesting in the above-mentioned fields. Samples of researches that have applied time series techniques and models are presented in Table 4.1.

Two major applications are usually followed up by the time series modeling:

- Forecasting
- Synthetic data generation

Forecasting is the process of making statements about events whose actual outcomes have not yet been observed. A commonplace example might be estimation for some variable of interest at some specified future date. Synthetic data generation is any production of data applicable to a given situation that are not obtained by direct measurement.

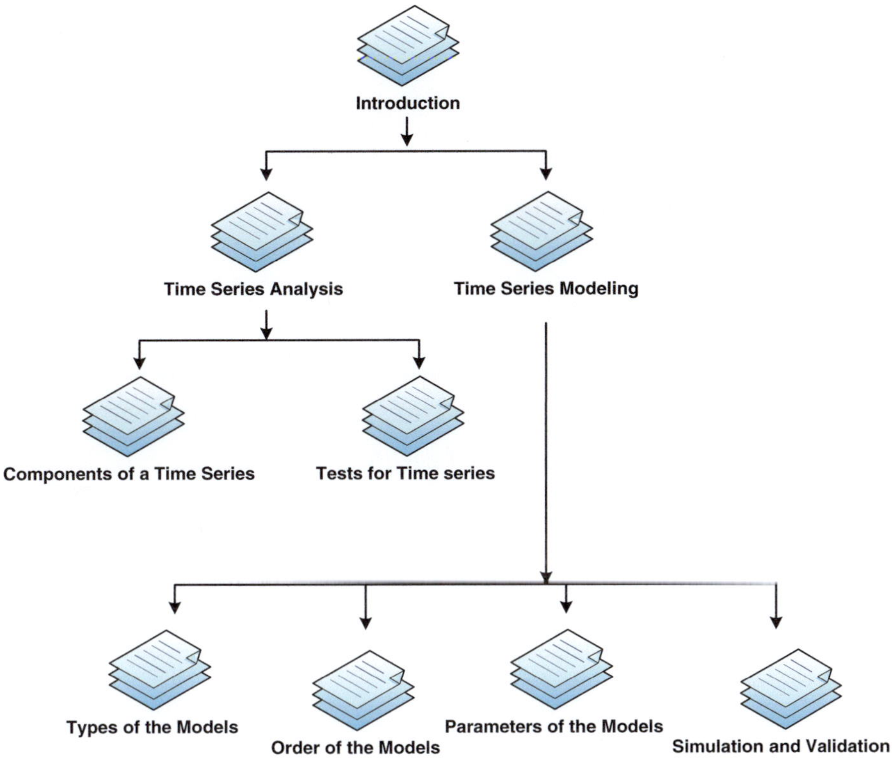

Fig. 4.1 Structure of the contents of this chapter

The applications of time series analysis and models expected to respond the following typical questions:

- How could we detect an increasing or decreasing trend in a time series?
- How could we detect a significant change in the long-term average of a time series?
- What would be the next streamflow data in near future, in case of considering a persistency within the correlation structure of the data?
- For designing purposes, how can we generate different realizations from an original long-term recorded data?

This chapter contains two major topics as shown in Fig. 4.1: "time series analysis" and "time series modeling." Time series analysis deals with components of a time series data as well as the common tests for time series which are usually applied to investigate their behavior before developing a model. Those tests might be useful to determine the appropriate approach for modeling a time series. The next topic, time series modeling, discusses time series models and how to apply them into the time series data. The topic includes models such as ARMA, ARIMA, and ARMAX. The application of the fitted models in time series forecasting, synthetic data generation, and multivariate modeling of time series are other addressed subjects. The chapter ends with a workshop.

4.2 Time Series Analysis

4.2.1 Components of a Time Series

To better define the components of a time series and analyze its behavior, some technical terms are explained before going through the next sections.

4.2.1.1 Periodicity

A function which output contains values that repeat periodically. In terms of time series modeling, the periodicity could be presented by the following relationship:

$$P_t = a\,\mathrm{Cos}\,(2\pi t/T) + b\,\mathrm{Sin}\,(2\pi t/T) \tag{4.1}$$

where P_t = periodical time series data, t = time as the independent variable of the equation which is the dynamic input of the equation, and T = return period of a whole cycle of time series which along with a and b are the parameters of the model. Periodicity is usually discovered in seasonal time series such as seasonal rainfall and air temperature data. So, this term might be frequently used in modeling of water resources and environmental systems.

4.2.1.2 Trend

A particular ascending or decreasing direction in the time series data. The linear trend is presented by the following equation:

$$Tr_t = at + b \tag{4.2}$$

where Tr_t = linear trend, t = time as the independent variable of the equation, and a and b are =parameters of the model. The detection of trend in time series of natural variables has become of more significance in recent decades due to the climate change phenomena.

4.2.1.3 Random Variable

A random variable is a variable whose value is subject to variations due to chance. As opposed to other mathematical variables, a random variable conceptually does not have a single, fixed value; rather, it can take on a set of probable different values, each with an associated probability. The interpretation of a random variable depends on the interpretation of its probability distribution function. This term is the major reason that makes the process of time series modeling a stochastic process.

4.2.1.4 Normal Standard Random Variable

A random variable with the normal standard distribution function.

4.2.1.5 Stochastic Time Series

A stochastic time series, Z_t, is presented in the general form of

$$Z_t = D_t + \varepsilon_t \tag{4.3}$$

where D_t is a deterministic regression-based mathematical function and ε_t is the normal standard random variable.

4.2.1.6 Time Series Modeling

Modeling a real-world stochastic time series, X_t, which might be a combination of the following terms:

$$X_t = P_t + Tr_t + Z_t \tag{4.4}$$

where P_t = deterministic periodic term, Tr_t = deterministic linear or nonlinear trend, and Z_t = stochastic term as discussed in Eq. (4.3).

4.2.1.7 Annual and Seasonal Time Series

The term annual time series refers to those types of time series which are presented in annual time intervals. Seasonal time series are those types of time series which are repeated periodically, usually in monthly and seasonal return periods. A seasonal effect is a systematic and calendar-related effect. Seasonal adjustment is the process of estimating and then removing the seasonal behavior from a time series.

4.2.1.8 Jump

A sudden significant change observed in the long-term average of a time series. A time series with *jump* is not stationary type I since its average is not constant in the entire length of the time series. Figure 4.2 shows a time series which includes a jump in time 25. It is to be notified that the fluctuation seen in time 13 or others except 25 is not considered as a "*jump*" because no significant difference in the average of data exists before and after that time.

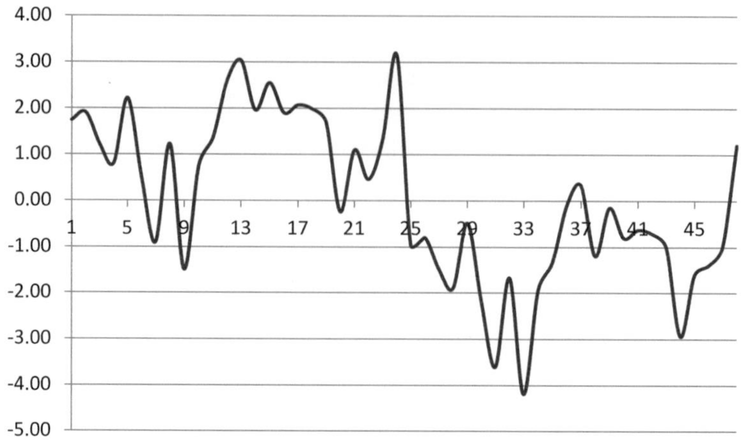

Fig. 4.2 A time series with a jump

Figure 4.3a demonstrates an example of a time series data. This data is in fact a combination of four major components including a periodic fluctuation as $P_t = 0.9\ \mathrm{Cos}(2\pi t/12) + \mathrm{Sin}(2\pi t/12)$ shown in Fig. 4.3b, a linear ascending trend as $Tr_t = 0.09t + 0.02$ shown in Fig. 4.3c, a correlated data as $x_t = 0.9x_{t-1}$ shown in Fig. 4.3d, and finally random variables shown in Fig. 4.3e.

The data shown in Fig. 4.3 are tabulated in Table 4.2 for more details.

4.2.2 *Tests for Time Series*

Preprocessing of a time series involves exploring periodicity, trend, and correlation between the data. This section presents two tests for a time series, namely, *Mann–Kendall* test of trend and *Kruskal and Wallis* test of jump.

The general procedure for the tests presented in this chapter is summarized by the following algorithmic steps:

- Chose a *null hypothesis*, H_0.
- Chose an *alternative hypothesis*, H_1.
- Choose a *significance level*, α (usually 0.05).
- Compute a *test statistic* using sample data.
- Compute *critical values* which are usually obtained from the statistical tables.
- *Compare* the computed test statistic with the critical values.
- If computed statistic lies within the critical limits, then accept H_0, otherwise, reject H_0 and accept H_1.

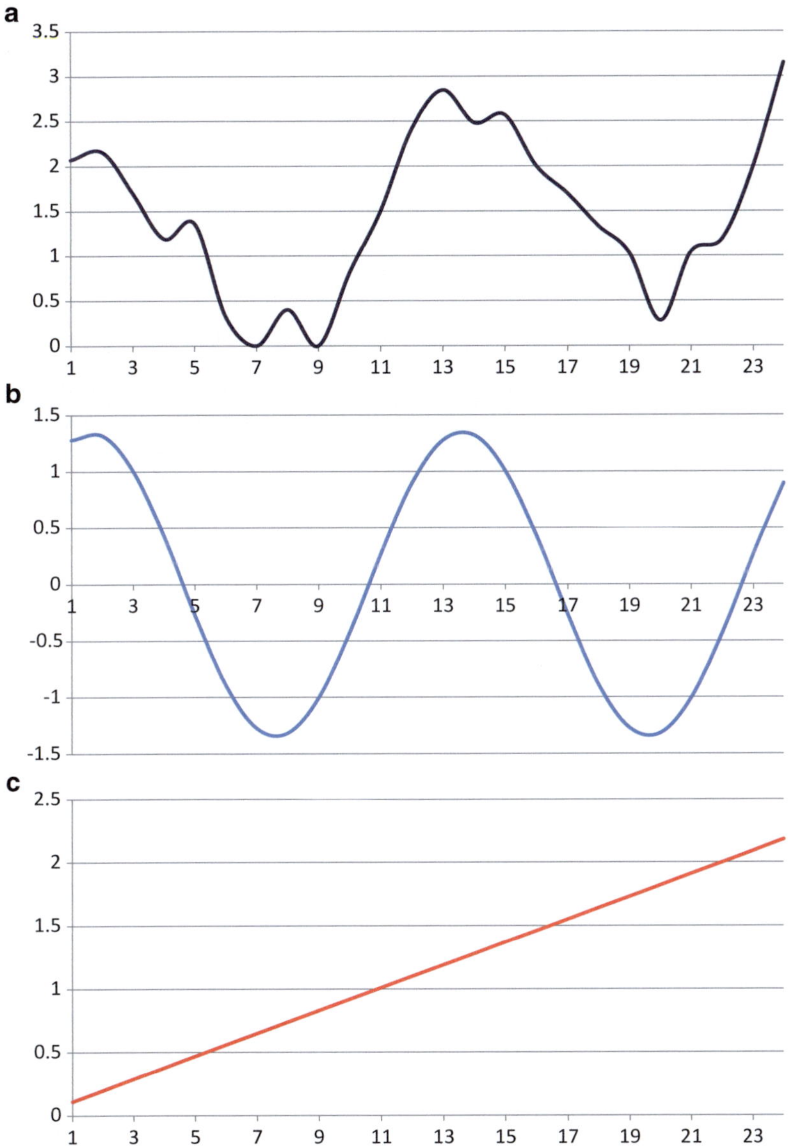

Fig. 4.3 Different components of a time series (**a**): periodic term (**b**), linear trend (**c**), regressive model (**d**), and random variables (**e**)

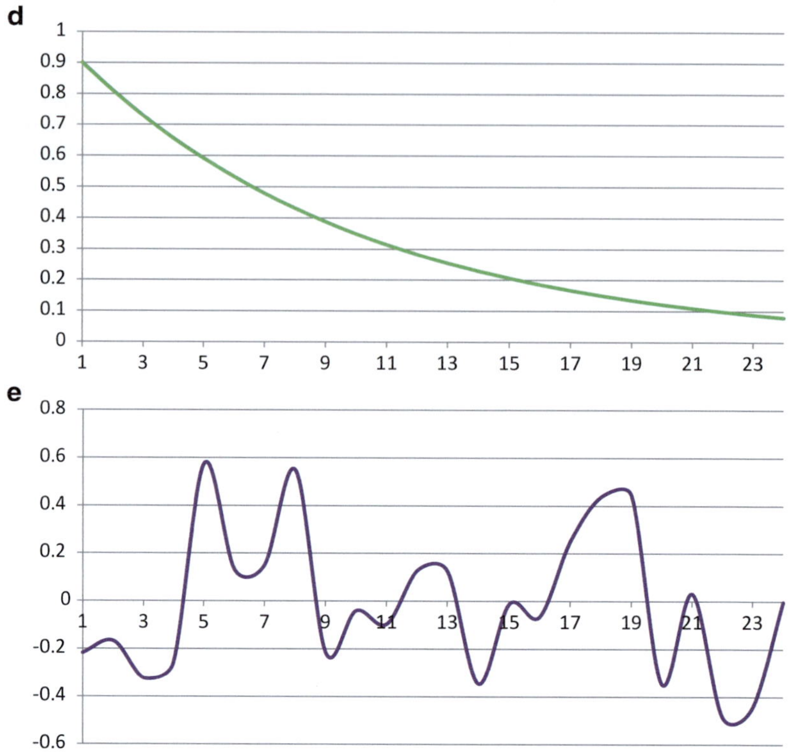

Fig. 4.3 (continued)

Table 4.2 An example of a time series and its components

No.	Original time series	Periodic term	Linear trend	Autocorrelated data	Random variables
1	2.070912	1.279312	0.11	0.9	−0.2184
2	2.158174	1.316174	0.2	0.81	−0.168
3	1.695917	1.000716	0.29	0.729	−0.3238
4	1.191884	0.417384	0.38	0.6561	−0.2616
5	1.358166	−0.27768	0.47	0.59049	0.575352
6	0.323111	−0.89841	0.56	0.531441	0.130076
7	0	−1.27865	0.65	0.478297	0.150351
8	0.401971	−1.31662	0.74	0.430467	0.548119
9	0	−1.00215	0.83	0.38742	−0.21527
10	0.808361	−0.41942	0.92	0.348678	−0.0409
11	1.502811	0.275579	1.01	0.313811	−0.09658
12	2.406922	0.89681	1.1	0.28243	0.127683
13	2.84731	1.27798	1.19	0.254187	0.125143
14	2.479871	1.317055	1.28	0.228768	−0.34595
15	2.567446	1.003576	1.37	0.205891	−0.01202

(continued)

Table 4.2 (continued)

No.	Original time series	Periodic term	Linear trend	Autocorrelated data	Random variables
16	2.000806	0.421455	1.46	0.185302	−0.06595
17	1.694373	−0.27348	1.55	0.166772	0.251083
18	1.332189	−0.89521	1.64	0.150095	0.437306
19	1.031486	−1.27731	1.73	0.135085	0.443709
20	0.278625	−1.31749	1.82	0.121577	−0.34546
21	1.045361	−1.005	1.91	0.109419	0.030944
22	1.189341	−0.42349	2	0.098477	−0.48565
23	2.004612	0.271383	2.09	0.088629	−0.4454
24	3.150638	0.893611	2.18	0.079766	−0.00274

4.2.2.1 *Mann–Kendall* Test of Trend

Mann–Kendall test has been presented by Mann (1945) and extended by Kendall (1975). This test is applied to check whether a time series contains an increasing or a decreasing trend. If a time series contains a trend, its trend component should be modeled in the process of time series modeling. *Mann–Kendall* test uses the general predefined procedure of the hypothesis test presented before. Every hypothesis test requires the analyst to state a null hypothesis and an alternative hypothesis. The hypotheses are stated in such a way that they are mutually exclusive. That is, if one is true, the other must be false, and vice versa. For a *Mann–Kendall* test of trend, the hypotheses take the following form:

Null hypothesis, H_0: time series has no trend.
Alternative hypothesis, H_1: time series has either an increasing or a decreasing trend.
Test statistic: the test statistic, Z, is obtained as follows:

First, for a time series, X_i, S is calculated as

$$S = \sum_{k=1}^{n-1} \sum_{j=k+1}^{n} \operatorname{sgn}(x_j - x_k) \tag{4.5}$$

where

$$\operatorname{sgn}(x) = \begin{cases} +1 & \text{if} \quad (x_j - x_k) > 0 \\ 0 & \text{if} \quad (x_j - x_k) = 0 \\ -1 & \text{if} \quad (x_j - x_k) < 0 \end{cases} \tag{4.6}$$

In case that the data is independent and uniformly distributed, the average of S is

$$E(S) = 0 \tag{4.7}$$

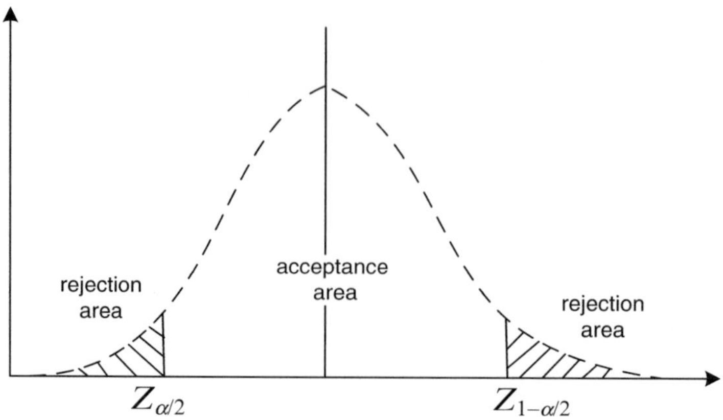

Fig. 4.4 The area of acceptance and rejection of null hypothesis in *Mann–Kendall* test

and its variance will be

$$\mathrm{Var}(S) = \frac{n(n-1)(2n+5) - \sum_{i=1}^{m} t_i(t_i-1)(2t_i+5)}{18} \tag{4.8}$$

In the above equation, n is the number of data and m is the number of ties. Each tie is a set of similar consequent data in a time series where number of data in each of them is t.

Finally, the statistic of this test, Z, is computed as

$$Z = \begin{cases} \dfrac{S-1}{\sqrt{\mathrm{Var}(S)}} & \textit{if} \quad S > 0 \\ 0 & \textit{if} \quad S = 0 \\ \dfrac{S+1}{\sqrt{\mathrm{Var}(S)}} & \textit{if} \quad S < 0 \end{cases} \tag{4.9}$$

Critical values: *Mann–Kendall* is a two-tailed test, which means that the test statistic should be compared by a critical value from the table of normal standard distribution, $Z_{\alpha/2}$, as shown in Fig. 4.4. In case that $Z_{\alpha/2} \leq Z \leq Z_{1-\alpha/2}$, the null hypothesis is accepted by α percent of error. In case that the null hypothesis is rejected, the time series of X_i is considered to have either an increasing trend if S is positive or a decreasing trend if S is negative.

The following program applies *Mann–Kendall* test for a time series X_i:

```
%% the program for Mann-Kendall test

sizeN=size(x)
N=input('please enter N');

%% calculation of S

S=0;
for k=1:N-1
    sumation(k)=0;
    num(k)=1;
    for l=k+1:N
    D=sign(x(l)-x(k));
    if D==0
      num(k)=num(k)+1;
    end
    sumation(k)=sumation(k)+D;
    end
    S=S+sumation(k);
end

%% calculation of number of ties

m=0;
for t=1:N-1
    if num(t)>1
    m=m+1;
    tie(m)=num(t);
    A=x(t);
    for r=t+1:N-1
      if x(r)==A
        num(r)=1;
      end
    end
  end
end

%% calculation of variance

sumtie=0;
for j=1:m
```

(continued)

```
    p=tie(j)*(tie(j)-1)*(2*tie(j)+5)/18;
    sumtie=sumtie+p;
end
var=N*(N-1)*(2*N+5)/18-sumtie;

%% calculation of Z and comparison with ns Z

if S>0
  Z=(S-1)/(var^0.5);
elseif S==0
  Z=0;
else
  Z=(S+1)/(var^0.5);
end
alpha=input('please enter alpha');
prob=1-alpha/2;
Zstandard=norminv(prob,0,1);
if abs(Z)<Zstandard
   fprintf('with probability of %f percent\n this time
series has NO significant TREND\n',(1-alpha)*100);
else if S>0
   fprintf('with probability of %f percent\n this time
series has an UPWARD TREND\n',(1-alpha)*100);
    else
   fprintf('with probability of %f percent\n this time
series has a DOWNWARD TREND\n',(1-alpha)*100);
  end
end
xx=x(1:N);
t=1:N;
plot(t,xx),title('time series x'),xlabel('t'),ylabel
('x'),grid;
```

Table 4.3 shows variables and parameters used in the program, which might be changed due to the change of the problem.

Example 4.1: Testing Trend in a Time Series

Test the possible trend in the two standardized rainfall time series recorded in a station and generated by synthetic data as presented in Table 4.4.

Mann–Kendall test is run for both time series with $N = 24$, alpha $= 0.05$, and x as the name of time series. The test demonstrates that the first time series has no significant trend and the second one has an increasing trend.

Table 4.3 Input data needed for the presented *Mann–Kendall* program

Variable/ parameter	Description
x	Name of the matrix of the specific time series
N	Number of data in the time series
alpha	Selected level of significance

Table 4.4 Data presented for Example 4.1

No.	First time series	Second time series
1	−0.55	0.11
2	−0.42	0.2
3	−0.81	0.29
4	−0.65	0.38
5	1.44	0.47
6	0.33	0.56
7	−0.75	0.65
8	1.37	0.74
9	−1.71	0.83
10	−0.10	0.92
11	−0.24	1.01
12	0.32	1.1
13	0.31	1.19
14	−0.86	1.28
15	−0.03	1.37
16	−0.16	1.46
17	0.63	1.55
18	1.09	1.64
19	1.11	1.73
20	−0.86	1.82
21	0.08	1.91
22	−1.21	2
23	−1.11	2.09
24	−0.01	2.18

4.2.2.2 *Kruskal* and *Wallis* Test of Jump

This test has been presented by *Kruskal* and *Wallis* in 1952 to check whether a time series contains a jump (Kruskal and Wallis 1952). In case of a jump, a time series is usually divided in two sets of data before and after the time of observation the *jump*. Jump occurs when land use change or interbasin water transfer causes significant changes to the natural water yield of a river. In those cases detecting a jump prevents misleading about the real potential of a river.

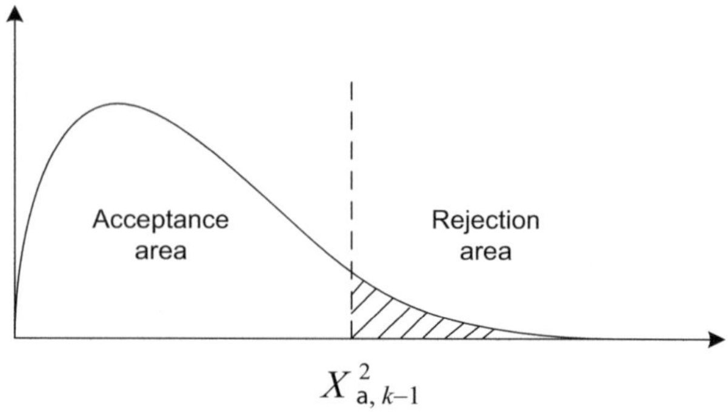

Fig. 4.5 The area of acceptance and rejection of null hypothesis in *Kruskal–Wallis* test

The following specific definitions are used through the procedure of this test:

H_0: No jump is observed in the time series.
H_1: The time series contains a jump.

Test statistic: the statistic of this test is H, which is calculated as follows.

A time series X_i is divided into k groups of data where each group j contains n_j data. Each jump divides the time series into two groups. The order of data within each group (r_{ij}) is calculated and R_j is obtained as

$$R_j = \sum_{i=1}^{n_j} r_{ij} \quad j = 1, 2, \ldots, k \tag{4.10}$$

Similar data get similar orders. The statistic of this test is H, which is calculated as

$$H = \frac{12}{N(N+1)} \sum_{j=1}^{k} \left(\frac{R_j^2}{n_j} \right) - 3(N+1) \tag{4.11}$$

where $N = \sum_{j=1}^{k} n_j$ and

Critical values: The above statistic will have a chi-square distribution with $k - 1$ degree of freedom (k is the number of data). In case that $H \leq X^2_{\alpha,(k-1)}$, the null hypothesis is accepted (Fig. 4.5).

The following program applies *Kruskal–Wallis* test for a time series X_i:

```
% Kruscal-Wallis test of JUMP

sizeN=size(x)
N=input('please enter N');
```

(continued)

```
% calculation of R(x) = rank of any x
y=sort(x);
grade(1) = 1;
for i = 2:N
  if y(i)>y(i-1)
    grade(i) = grade(i-1)+1;
  else
    grade(i) = grade(i-1);
  end
end
for i = 1:N
  for j = 1:N
    if x(i) == y(j)
      R(i) = grade(j);
    end
  end
end
% determining the jumping-point with drawing plot
t = 1:N;
plot(t,x),title('time  series  x'),xlabel('t'),ylabel
('X(t)'),grid;
pause

% calculation of H and comparison with chi2 distribution

k = input('please enter k=number of groups');
step(1) = 0;
for j = 2:k+1
   step(j) = input('please enter step(j)=t at the end of
group j');
  % step(j) = t at the end of group j
end
sumation = 0;
for j = 2:k+1
  sumR = 0;
  for l = step(j-1)+1:step(j)
    sumR = sumR+R(l);
  end
  s=sumR^2/(step(j)-step(j-1));
  sumation=sumation+s;
end
H = (12/(N*(N+1)))*sumation-3*(N+1);
```

(continued)

```
alpha = input('please enter alpha');
p = 1-alpha;
dof = k-1;
Hjchi2 = chi2inv(p,dof);
if H<Hjchi2
   fprintf('with probability of %f percent\n this time
series has NO significant JUMP\n',p*100);
else
   fprintf('with probability of %f percent\n this time
series HAS significant JUMP\n',p*100);
end
```

Table 4.5 demonstrates variables and parameters used in the program, which might be changed due to the change of the problem.

Example 4.2: Test of Jump in a Time Series

Investigate the existence of jump in the following time series which represents inflow to a reservoir (Table 4.6).

Solution
Time Series 1

Selecting $N = 48$ and $k = 3$, time series is divided to three groups of data from 1 to 14, 15 to 34, and 35 to 48. The groups are defined by entering 14, 34, and 48 as different $S(j)$ values, which is asked by the program. As it is demonstrated by the output of the program, time series 1 has no significant jump in the significance level of 0.05 (Fig. 4.6).

Time Series 2

Selecting $N = 48$ and $k = 2$, time series is divided to two groups of data from 1 to 23 and 24 to 48. The groups are defined by entering 23 and 48 as different $S(j)$ values, which is asked by the program. As it is demonstrated by the output of the program, time series 2 has a jump in the significance level of 0.05 (Fig. 4.6).

Table 4.5 Input data needed for the presented test of *Jump* program

Variable/parameter	Description
x	Name of the time series of a specific problem
k	Number of groups
N	Number of data in the time series of the problem
Step(*j*)	Time step when the last data of group *j* has been recorded
alpha	Significance level

Table 4.6 Data presented for Example 4.2

No.	Series 1	Series 2	No.	Series 1	Series 2
1	1,765.0	1.74	26	2,834.6	−0.79
2	3,212.0	1.91	27	2,492.6	−1.49
3	2,341.0	1.21	28	2,972.4	−1.90
4	2,141.5	0.80	29	2,946.2	−0.48
5	2,341.0	2.22	30	3,106.6	−2.18
6	2,280.0	0.52	31	4,994.0	−3.61
7	2,946.2	−0.91	32	3,128.0	−1.67
8	1,722.3	1.22	33	2,929.2	−4.20
9	2,345.0	−1.50	34	1,752.3	−1.95
10	1,316.9	0.75	35	5,964.4	−1.34
11	2,679.5	1.36	36	2,946.2	−0.10
12	3,249.0	2.60	37	3,476.3	0.34
13	3,208.7	3.04	38	5,291.3	−1.20
14	3,014.3	1.96	39	4,023.6	−0.15
15	5,681.5	2.55	40	2,854.0	−0.80
16	3,473.5	1.90	41	3,983.3	−0.63
17	5,463.0	2.07	42	1,757.0	−0.71
18	3,368.5	1.99	43	1,103.5	−1.00
19	5,425.2	1.70	44	847.5	−2.94
20	2,511.8	−0.24	45	3,212.0	−1.61
21	2,770.8	1.09	46	2,236.6	−1.40
22	2,137.1	0.46	47	2,341.8	−1.00
23	5,226.6	1.34	48	2,112.6	1.20
24	3,573.2	3.15	49		
25	2,888.5	−0.96	50		

4.3 Time Series Models

Time series analysis accounts for the fact that data points taken over time may have an internal structure (such as autocorrelation, trend, or seasonal variation) that should be modeled. This section will give a brief overview of some of the more widely used techniques which are used for time series modeling.

The process of time series modeling consists of four major steps of

- Model selection
- Selection of the order of models
- Determining the model parameters
- Simulation and validation

which are defined as follows:

4.3.1 Model Selection

The characteristics of a time series determine which model fits better to reflect its variation. Different models such as autoregressive (AR), auto Regressive moving

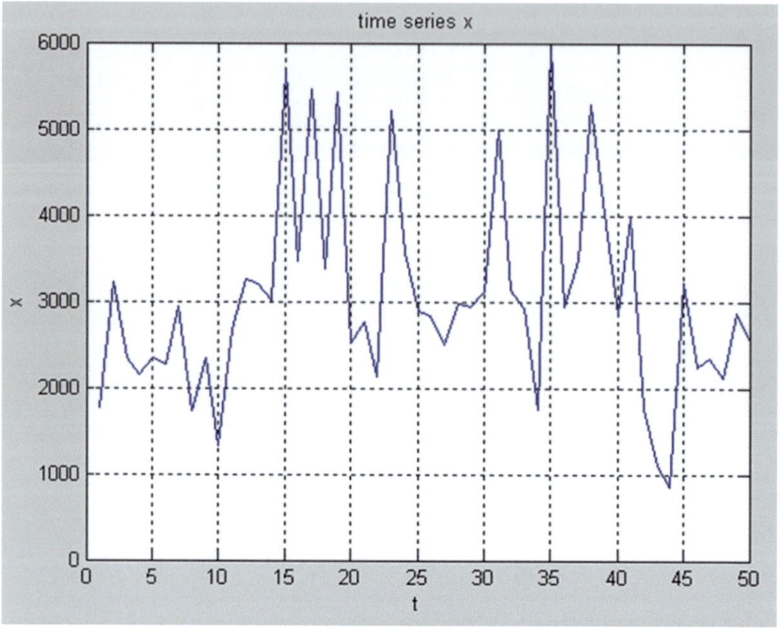

Time series 1

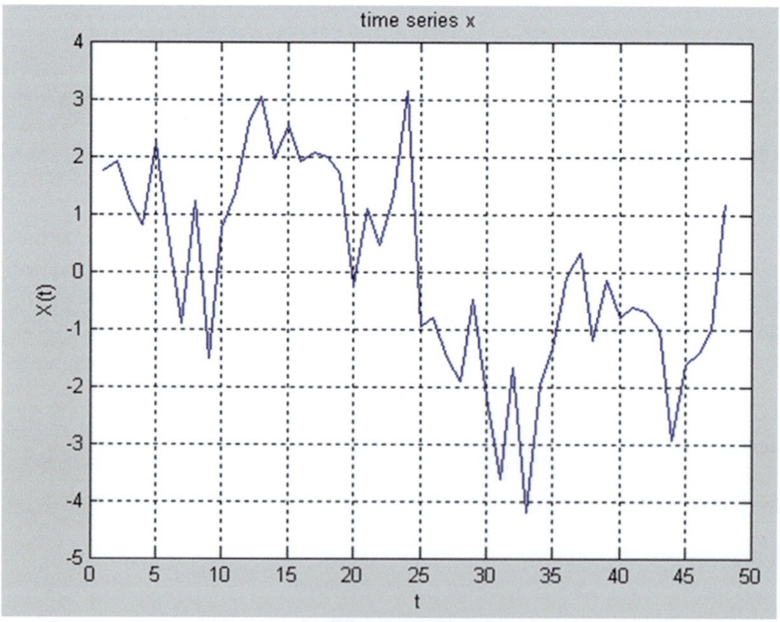

Time series 2

Fig. 4.6 Time series of Example 4.2

average (ARMA), and autoregressive integrated moving average (ARIMA) can be considered as alternatives of a time series model.

4.3.2 Order of Models

Each of the above models represents a wide range of variants with different orders. In addition to the type of the model, the order of the model should be determined too. For example, an ARMA model is the general term for an ARMA(p,q) model. The values of p and q determine the exact order of the extended equation that an ARMA model is representing. The order of the model usually determines the short-term and long-term memory within a time series.

4.3.3 Determining the Model Parameters

Different parameters of the model are calculated in this step. The number of parameters that a model might have is related to both its type and its order.

4.3.4 Simulation and Validation

The aim of time series modeling is to simulate the variation of time series for generating and/or forecasting data. After data simulation, the skill of each selected model should be evaluated using different criteria.

4.3.5 Types of Time Series Models

4.3.5.1 Autoregressive, AR(p), Models

Since the 1960s, autoregressive models have been widely used in water resources engineering. The basis for *AR* models is the *Markov* chain, which relates to the time series where every individual data is correlated with its past and future data. A *Markov* chain is a mathematical system that undergoes transitions from one state to another, between a finite or countable number of possible states (Salas et al. 1980).

The basic idea for an autoregressive model comes from the idea of a linear regression model as $Y = \beta_1 X + \beta_0 + \varepsilon$, where Y is the dependent variable, X is the independent variable, and ε is a random number. In case of time series modeling, the dependent and independent variables come from a similar data. It means that both of them refer to an original time series. Here, the difference between independent and dependent variables comes from the difference between the time of occurrence of each data or in fact the index of data. Since the value of data in a time series depends on the time of recording the data, the following equation could be an alternative of the above linear equation:

$$X_t = \sum_{i=1}^{k} \varphi_i X_{t-i} + \varepsilon_t \tag{4.12}$$

In the above model X = the time series data, k = lag time between the current variable and the recorded past data, and φ_i = parameters of the linear regression. This equation is known as the autoregressive model.

The term $\sum_{i=1}^{k} \varphi_i X_{t-i}$ represents the deterministic component of the stochastic variable, X, and ε_t is the random component of it. The random term of the above equation is considered as a standard normal random variable. For the compatibility of the random and deterministic components of the above equation, X should be transformed to a normal standard variable by the following procedure:

$$Y_t = g(X_t) \tag{4.13}$$

where g is a transformation function, which transforms a data of specific distribution to the data of normal distribution. The normalized data is then standardized by

$$Z_t = \frac{Y_t - \mu}{\sigma} \tag{4.14}$$

where μ and σ = the average and standard deviation of Y_t time series, respectively. One of the alternatives for the "g" transformation function is the *Box–Cox* function (Box and Cox 1964) as

$$Y_t = \begin{cases} \dfrac{(X_t)^\lambda - 1}{\lambda} & if \quad \lambda \neq 0 \\ \log(X_t) & if \quad \lambda = 0 \end{cases} \tag{4.15}$$

where λ = the parameter of the transformation.

The optimum λ parameter for a specific time series (data) can be obtained by the following syntax:

```
[Y, lambda] =boxcox(data)
```

The normal standard variable is then calculated by

```
Z=(Y-mean(Y))/std(Y)
```

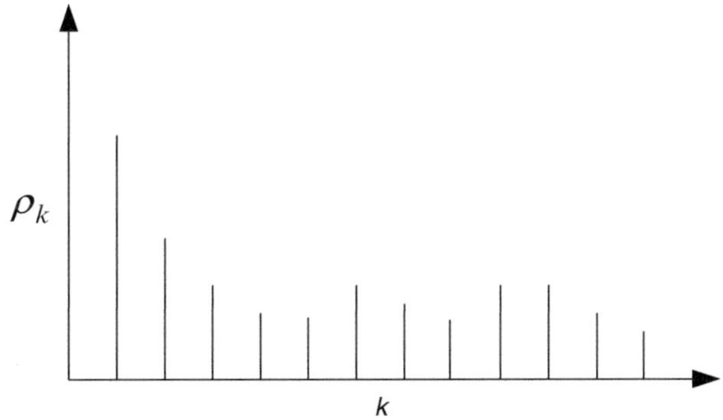

Fig. 4.7 Schematic of a correlogram

And Eq. (4.12) is applied to Z time series as

$$Z_t = \sum_{i=1}^{p} \varphi_i Z_{t-i} + \varepsilon_t \tag{4.16}$$

Similar to the well-known linear regression models, a correlation coefficient can be used to represent how correlated the data of a time series are. The following equation shows the formulation of this correlation coefficient:

$$\rho_k = \frac{\sum_{t=1}^{N-k} (Z_t - \mu)(Z_{t+k} - \mu)}{\sum_{t=1}^{N} (Z_t - \mu)^2} \tag{4.17}$$

where Z_t = the time series data, N = number of time series data, k = lag time between data, and μ = mean of the time series data.

The above coefficient, which shows the correlation between the data of a time series, is known as the autocorrelation function (ACF). In contrast to the ordinary correlation coefficient, the autocorrelation function gets different values for different lag time, k. So, to better represent the autocorrelation function of a time series, different ρ_k values are plotted versus different lag times, k. This plot, which is known as "*correlogram*," is usually used to define the correlation between time series data and consequently the order of the model (which is going to be discussed in the next section). A schematic of a correlogram is shown in Fig. 4.7. Significant values of ρ could help choosing the appropriate parameters for the time series of Eq. (4.16).

4.3.5.2 Autoregressive Moving Average, ARMA(*p*,*q*), Model

The experience of applying AR(p) models to the real-world data might result in significant simulation errors. The reason for this is related to the fact that the occurrence of Z_t in the current time is only considered as a function of its previous values. It is a fact that the effect of other predictors are neglected in the conventional AR(p). Suppose that we are modeling the streamflow of a river by an AR(p) model. Obviously, the streamflow is a function of the base flow, snow budget, soil moisture, etc., which are not considered in the *AR* model at all. A solution to mitigate this shortcoming of the *AR* models is to consider the error term, ε_t, as a representative of all other predictors except the persistency of the time series data.

The AR(p) model assumes that the stochastic variable Z_t is consisted in the form of a deterministic term as $\sum_{i=1}^{p} \varphi_i Z_{t-i}$ and a random term as ε_t . ε_t can be considered as a stochastic variable instead of a random variable. The stochastic variable, ε_t, can be then represented by an autoregressive model itself as

$$\varepsilon_t = \sum_{j=1}^{q} \theta_j \varepsilon_{t-j} + \zeta_t \tag{4.18}$$

where θ_j = parameters of the model and ζ = random variable or the error of the model. The above model is known as the moving average model of order q, MA(q).

Now, after subtracting the modeled error from Eq. (4.16), the AR model can be extended to the following form:

$$Z_t = \sum_{i=1}^{p} \varphi_i Z_{t-i} - \varepsilon_t = \sum_{i=1}^{p} \varphi_i Z_{t-i} - \sum_{j=1}^{q} \theta_j \varepsilon_{t-j} + \zeta_t \tag{4.19}$$

The above equation is known as the autoregressive moving average of orders p and q, ARMA(p,q).

4.3.5.3 Autoregressive Integrated Moving Average, ARIMA(*p*,*d*,*q*), Model

An autoregressive integrated moving average (ARIMA) model is a generalization of an autoregressive moving average (ARMA) model. These models are fitted to time series data either to better understand the data or to predict future points in the series (forecasting). They are applied in some cases where data are not stationary type I and an initial differencing step (corresponding to the "integrated" part of the model) can be applied to remove the trend of the data.

The model is generally referred to as an ARIMA(p,d,q) model where p, d, and q are nonnegative integers that refer to the order of the autoregressive, integrated, and moving average parts of the model, respectively. p and q refer to the order of ARMA model and d refers to the order of differencing.

A linear trend in a time series, T_t, can be represented by $T_t = at + b$ where t is the time step and a and b are the constant parameters. The trend is changed to a constant term by a first derivation of T_t over t. An alternative for the derivation is the gradient of $\frac{\Delta T}{\Delta t}$ or in case of constant time interval, ΔT. As an example, the first order (U) and second order difference (W) of time series Z are represented as

Table 4.7 Data presented for Example 4.3

Row	Data
1	10
2	11
3	17
4	18
5	20
6	25
7	24
8	27
9	25
10	32

Table 4.8 Results obtained by differencing of time series presented in Example 7.3

t	Z_t	U_t
1	10	
2	11	1
3	17	6
4	18	1
5	20	2
6	25	5
7	24	−1
8	27	3
9	25	−2
10	32	7

$$\begin{aligned} U_t &= Z_t - Z_{t-1} \\ W_t &= U_t - U_{t-1} \end{aligned} \tag{4.20}$$

The general equation of an ARIMA (p,d,q) model is then represented as

$$V_t = \sum_{i=1}^{p} \varphi_i V_{t-i} - \sum_{j=1}^{q} \theta_j \varepsilon_{t-j} + \zeta_t \tag{4.21}$$

where V_t is resulted by the dth difference of the original Z_t data through the preprocessing operation.

Example 4.3: The Difference Operator

Table 4.7 shows a time series with a linear trend. Use a difference operator to make the data ready for modeling.

Solution

The first order difference of the data is shown in Table 4.8.

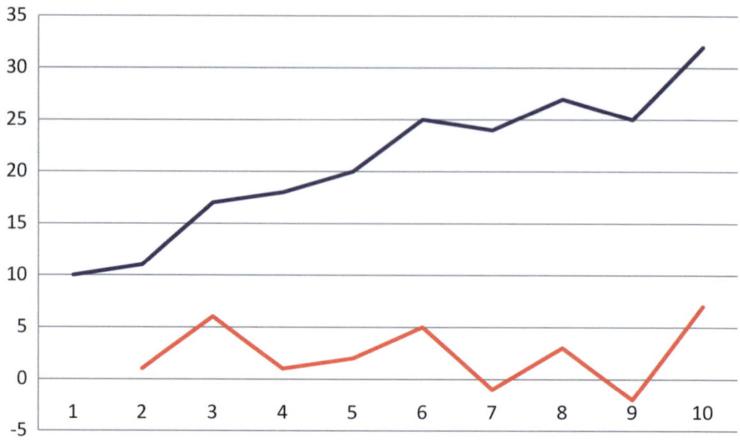

Fig. 4.8 Original and differenced time series of Example 4.3

And Fig. 4.8 shows the original time series as well as the differenced one which is actually a stationary data. It is to be notified that if the differenced data is still nonstationary, a higher order of differencing is necessary to process the data.

4.3.5.4 Autoregressive Moving Average Model with Exogenous, ARMAX, Inputs Model

To consider the effect of an external predictor except the persistence between data, an improved model of ARMA, named ARMAX (AutoRegressive Moving Average with eXogenous inputs), has been developed. In general, the *ARMAX*(p,q,b) model might be expected to simulate very well the behavior of systems whose input–output characteristics are approximately linear. The notation *ARMAX*(p,q,b) refers to the model with p autoregressive terms, q moving average terms, and b exogenous inputs terms. This model contains the AR(p) and MA(q) models and a linear combination of the last b terms of a known and external time series, X_t. It is given by

$$Z_t = \sum_{i=1}^{p} \varphi_i Z_{t-i} - \sum_{j=1}^{q} \theta_j \varepsilon_{t-j} + \sum_{l=1}^{b} \gamma_i X_{t-l} + \zeta_t \tag{4.22}$$

where Z_t = the dependent variable, X_t = the exogenous input, ε_t = the previous error of model, φ, θ, and γ = parameters of the model, p, q, and b = orders of the model, and finally ζ_t = modeling error.

4.3.5.5 Multivariate Time Series Modeling

Multivariate time series analysis is used when one wants to model and explain the interactions and co-movements among a group of time series variables. In many

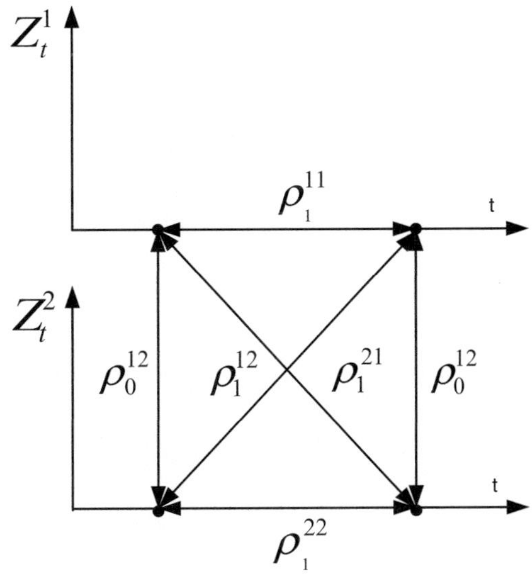

Fig. 4.9 The structure of autocorrelation and cross-correlation between two time series

problems of water resources and environmental engineering, it is essential to model two or more time series, simultaneously. It is because of the correlation between multiple variables. Examples of this type of modeling are dealing with the rainfall data of adjacent stations, rainfall and streamflow in a basin, pollutant variables measured along with a river, etc.

In multivariate modeling of time series, each component has autocovariances and autocorrelations, but there is also a cross-correlation between them which should be dealt with by the following equation:

$$\rho_k^{ij} = \frac{\sum_1^{n-k}\left(Z_t^i - \overline{Z}^i\right)\left(Z_{t+k}^j - \overline{Z}^j\right)}{\sum_1^{n}\left(Z_t^i - \overline{Z}^i\right)\sum_1^{n}\left(Z_t^j - \overline{Z}^j\right)} \tag{4.23}$$

where Z_t^i = ith time series, $\overline{Z}^i$ = the long-term average of the ith time series, and n = number of data in time series.

Figure 4.9 presents the structure of 1-lag correlation between two different time series.

The multivariate form of an AR(1) is

$$Z_t = A_1 Z_{t-1} + B\varepsilon_t \tag{4.24}$$

where Z_t is an $(n \times 1)$ vector, containing elements of n different time series, z_t^i. The extended form of Eq. (4.25) would then be

$$\begin{bmatrix} z_t^1 \\ z_t^2 \\ \vdots \\ z_t^n \end{bmatrix} = \begin{bmatrix} a^{11} & a^{12} & \dots & a^{1n} \\ a^{21} & a^{22} & \dots & a^{2n} \\ \vdots & \vdots & \vdots & \vdots \\ a^{n1} & a^{n2} & \dots & a^{nn} \end{bmatrix} \begin{bmatrix} z_{t-1}^1 \\ z_{t-1}^1 \\ \vdots \\ z_{t-1}^1 \end{bmatrix} + \begin{bmatrix} b^{11} & b^{12} & \dots & b^{1n} \\ b^{21} & b^{22} & \dots & b^{2n} \\ \vdots & \vdots & \vdots & \vdots \\ b^{n1} & b^{n2} & \dots & b^{nn} \end{bmatrix} \begin{bmatrix} \varepsilon_t^1 \\ \varepsilon_t^2 \\ \vdots \\ \varepsilon_t^n \end{bmatrix} \tag{4.25}$$

4.3.6 Order of Time Series Models

4.3.6.1 Order of AR(p) Models

Suppose that the best-fitted model on a normal standard time series is presented as

$$Z_t = \sum_{i=1}^{p} \varphi_i Z_{t-i} + \varepsilon_t \tag{4.26}$$

or

$$Z_t = \varphi_1 Z_{t-1} + \varphi_2 Z_{t-2} + \cdots + \varphi_p Z_{t-p} + \varepsilon_t \tag{4.27}$$

where Z_t is the normal standard data of the original time series of X_t which is obtained as $Z_t = \frac{Y_t - \mu}{\sigma}$, where μ and σ are the average and standard deviation of Y_t normal time series. Equation (4.26) could be rewritten as

$$\frac{Y_t - \mu}{\sigma} = \sum_{i=1}^{p} \varphi_i \left(\frac{Y_{t-i} - \mu}{\sigma} \right) + \varepsilon_t \tag{4.28}$$

The term $\frac{Y_{t-k} - \mu}{\sigma}$ is multiplied to both sides of the above equations.

$$\left[\frac{(Y_{t-k} - \mu)(Y_t - \mu)}{\sigma^2} \right] = \sum_{i=1}^{p} \varphi_i \left[\frac{(Y_{t-k} - \mu)(Y_{t-i} - \mu)}{\sigma^2} \right] + \varepsilon_t \tag{4.29}$$

Taking the expected value of each term of the above equation, the following equation resulted which is known as the Yule–Walker equation (Yule 1927; Walker 1931):

$$\rho_k = \sum_{i=1}^{P} \varphi_i \rho_{k-i} \quad k > 0 \tag{4.30}$$

or

$$\rho_k = \varphi_1\rho_{k-1} + \varphi_2\rho_{k-2} + \cdots + \varphi_p\rho_{k-p} \quad k > 0 \tag{4.31}$$

The above equation can be used to derive parameters of an AR(p) model substituting different values for p and k. For instance, for $k = 1$ and $p = 1$, the following equation is derived:

$$\rho_1 = \varphi_1 \tag{4.32}$$

and for $p = 2$ and $k = 1$

$$\rho_1 = \varphi_1/(1 - \varphi_2) \tag{4.33}$$

Among p numbers of φ, the pth parameter, φ_p, of an AR(p) is of more significance among the others. According to Eq. (4.31), the extent of the model is limited to the number of φ s. Obviously, φ parameters close to zero means that the correlation between that lag of time series is not significant. The order of an AR model is then limited to p. This is demonstrated by plotting partial autocorrelation (PAC) function, which is actually a plot of φ_p s versus different values of p. The following syntax is used to plot PACF for a time series:

```
parcorr(Z,Lag)
```

where Z is the time series and Lag is the lag time.

Example 4.4: Partial Autocorrelation Function

For the time series shown in Table 4.9, plot PACF and decide about the order of AR(p) model.

Solution
The PACF is plotted using the command of

```
parcorr(Z,Lag)
```

which results in the partial autocorrelation function shown in Fig. 4.10.

As it is obvious by the plot, PACF is almost zero from lag 2 to lag 8. Please note that the significant values of PACF from lag 9 later on might be a random

Table 4.9 Data presented for Example 4.4

No.	Time series data	No.	Time series data
1	119.1	14	97.0
2	107.1	15	87.3
3	96.4	16	78.6
4	101.0	17	87.0
5	90.9	18	78.3
6	81.8	19	110.0
7	73.6	20	65.0
8	66.3	21	65.0
9	59.6	22	58.5
10	122.0	23	52.6
11	109.8	24	47.4
12	98.8	25	78.0
13	88.9		

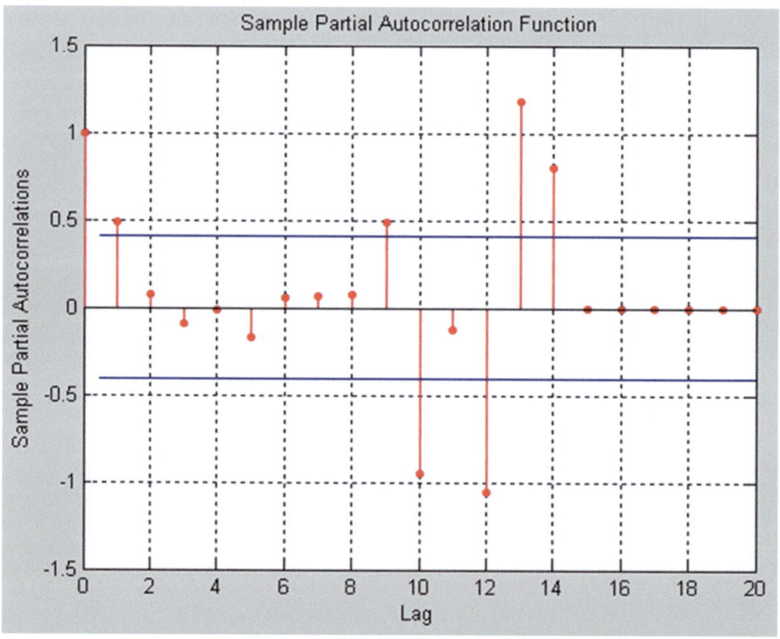

Fig. 4.10 Partial autocorrelation function for the time series of Example 4.4

phenomenon and does not mean any real correlation between data. As a result the optimum lag for an AR model to be fitted on the data is 1. It should be notified that the maximum lag time that the PACF could be drawn here is 24 (number of data, 1).

Table 4.10 Data presented for Example 4.5

No.	Time series data
1	1.83390
2	0.932676
3	1.158646
4	0.521661
5	0.150655
6	0.045185
7	0.349916
8	0.184503
9	0.56011
10	−1.06535

4.3.6.2 Order of MA(q) Models

From a procedure almost similar to deriving the Yule–Walker equation, order of an MA model is related to the autocorrelation coefficients which are obtained and plotted by the following command:

```
autocorr(Z,Lag)
```

ρ parameters close to zero mean that the correlation between that lag of residuals is not significant. The order of an MA model is then limited to significant values of ρ. This is demonstrated by plotting autocorrelation (AC) function, which is actually a plot of ρ s versus different values of q.

Example 4.5: Order of an MA(q) Model

For the normal standard time series shown in Table 4.10, decide on the order of the moving average model that might fit the data.

Solution

The autocorrelation of the above data is presented as follows (Fig. 4.11).

As it is obvious by the plot, ACF is almost zero from lag 3 later on. It is concluded that the best model for the data is MA(2).

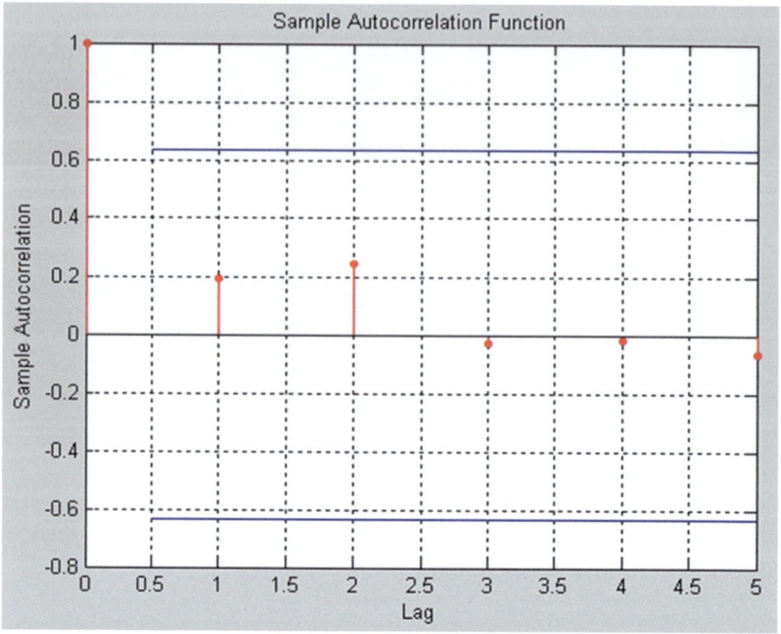

Fig. 4.11 Autocorrelation function for time series of Example 4.5

4.3.6.3 Order of ARMA(p,q) Models

Like AR and MA models, autocorrelation and partial autocorrelation functions can be used to determine the orders of an *ARMA*(p,q) model. The characteristic behavior of ACF and PACF based on Salas et al. (1980) is summarized as follows:

- For an ARMA(p,q) model, the autocorrelation function (ACF) is infinite in extent and the function is irregular in first $q - p$ lags then consists of damped exponentials and/or damped waves.
- For an ARMA(p,q) model, the partial autocorrelation function (ACF) is infinite in extent and the function is irregular in first $p - q$ lags then consists of damped exponentials and/or damped waves.
- For an AR(p) model, the partial autocorrelation function (PACF) is finite in extent with peaks at lags 1 through p then cuts off.
- For an AR(p) model, the autocorrelation function (ACF) is infinite in extent consisting of damped exponentials and/or damped waves.
- For an MA(q) model, the partial autocorrelation function (PACF) is infinite in extent consisting of damped exponentials and/or damped waves.
- For an MA(q) model, the autocorrelation function (ACF) is finite in extent with peaks at lags 1 through q then cuts off.

Table 4.11 Data presented for Example 4.6

Time	Data	Time	Data
1	1,765.0	26	2,834.6
2	3,212.0	27	2,492.6
3	2,341.0	28	2,972.4
4	2,141.5	29	2,946.2
5	2,341.0	30	3,106.6
6	2,280.0	31	4,994.0
7	2,946.2	32	3,128.0
8	1,722.3	33	2,929.2
9	2,345.0	34	1,752.3
10	1,316.9	35	5,964.4
11	2,679.5	36	2,946.2
12	3,249.0	37	3,476.3
13	3,208.7	38	5,291.3
14	3,014.3	39	4,023.6
15	5,681.5	40	2,854.0
16	3,473.5	41	3,983.3
17	5,463.0	42	1,757.0
18	3,368.5	43	1,103.5
19	5,425.2	44	847.5
20	2,511.8	45	3,212.0
21	2,770.8	46	2,236.6
22	2,137.1	47	2,341.8
23	5,226.6	48	2,112.6
24	3,573.2	49	2,871.6
25	2,888.5	50	2,543.9

Example 4.6: Order of an ARMA(p,q) Model

For the time series shown in Table 4.11, decide on the type and order of the model.

Solution

By plotting both ACF and PACF as presented in Figs. 4.12 and 4.13, we can decide on the type and order of the model. The plots demonstrate that an ARMA model is more suitable for the data since none of the plots have been cut off at a specific lag. Furthermore, the correlation by lag 2 is more significant than the others in both plots. So, ARMA(2,2) and lower orders are selected for the time series.

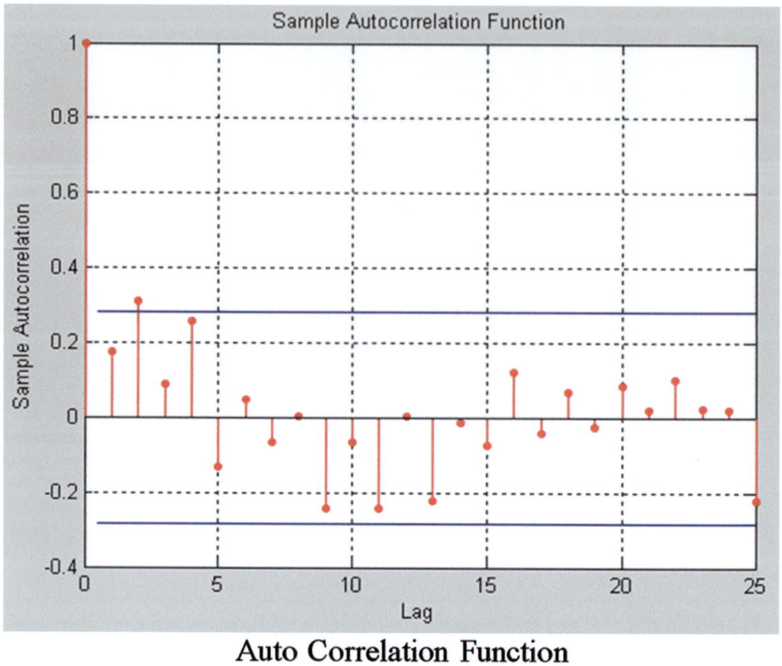

Fig. 4.12 ACF for the time series of Example 4.6

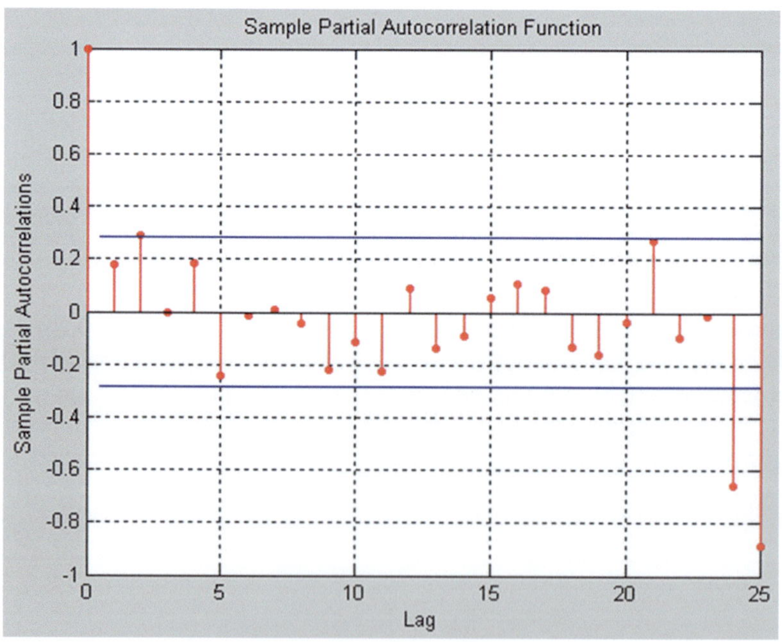

Fig. 4.13 PACF for the time series of Example 4.6

4.3.7 *Determining Parameters of the Time Series Models*

4.3.7.1 ARMAX Models

To determine the parameters of AR, ARMA, and ARMAX models in MATLAB, we need to consider the following general equation of ARMAX as follows to use armax command:

$$Z_t - \sum_{i=1}^{na} \varphi_i Z_{t-i} = \sum_{l=1}^{nb} \gamma_i X_{t-l-nk} - \sum_{j=1}^{nc} \theta_j \varepsilon_{t-j} + \xi_t \tag{4.34}$$

```
m= armax(Z,[na nb nc nk]);
```

In the armax command, Z is a two-column matrix containing the dependent variable in the first column and the exogenous variable in its second column. In case of modeling by AR and ARMA models, Z contains only one column of dependent variable. *na*, *nb*, *nc*, and *nk* are the orders and lags of the model as presented in Eq. (4.34). It is to be notified that in case of AR and ARMA models, *nb* and *nk* are removed from the command. Finally *m* contains the parameters of the model. It should be notified that the parameters of an ARIMA model are also obtained by the above-mentioned command. The difference is only in the preprocessing of time series which needs to be differenced before being used through the ARIMA model.

Example 4.7: Parameters of an ARMAX Model

Determine the parameters of an ARMA(1,2) and ARMAX(1,2,3) for the data shown in Table 4.12.

For ARMAX(1,2) we consider one-column matrix Z containing the original data and the parameters as

```
na=1;
nc=2;
```

The command is

```
m= armax(Z,[na nc]);
```

and the parameters will then be

Table 4.12 Time series presented for Example 4.7

Time	Z	X
1	17.65	18
2	32.12	30
3	23.41	28
4	21.42	22
5	23.41	22
6	22.8	31
7	29.46	32
8	17.22	33
9	23.45	34
10	13.17	35
11	26.8	36
12	27	27
13	32.09	38
14	30.14	39
15	56.82	40
16	34.74	41
17	44	43
18	33.69	42
19	54.25	50
20	25.12	45
21	27.71	43
22	21.37	44
23	52.27	24
24	35.73	49
25	28.89	50

```
A(q) = 1 -  1.022 q^-1

C(q) = 1 - 0.9515 q^-1 + 0.01017 q^-2
```

It means that φ_1, θ_1, and θ_2 are equal to -1.022, -0.9515, and 0.01017, respectively.

For ARMAX(1,2,3) we consider Z as a two-column matrix containing the dependent variable in the first column and the exogenous variable in its second column and

```
na=1;
nb=3;
nc=2;
nk=0;
```

The command is

```
m= armax(Z,[na nb nc nk]);
```

and the parameters will be

```
A(q) =1 - 0.6448 q^-1

B(q) =0.06291 + 0.1542 q^-1 + 0.1035 q^-2

C(q) =1 - 0.831 q^-1 - 0.1171 q^-2
```

It should be notified that transforming data to a normal standard series is necessary at the beginning of modeling in case of synthetic data generation.

4.3.7.2 Multivariate AR Models

The key factor in parameter estimation of a multivariate AR model is to calculate the correlation matrix, M, which is defined as follows:

$$M_k = \begin{bmatrix} \rho_k^{11} & \rho_k^{12} & \cdots & \rho_k^{1n} \\ \rho_k^{21} & \rho_k^{22} & \cdots & \rho_k^{21} \\ \vdots & & \cdots & \\ \rho_k^{n1} & \rho_k^{n2} & \cdots & \rho_k^{nn} \end{bmatrix} \tag{4.35}$$

where ρ_k^{ij} = the correlation coefficient as defined in equation 4.23.

For a MAR(1) model in form of

$$Z_t = A_1 Z_{t-1} + B\varepsilon_t \tag{4.36}$$

the matrix A is obtained by the following multivariate form of Yule–Walker equation (Salas et al. 1980):

$$M_k = A_1 M_{k-1} \quad k > 0 \tag{4.37}$$

or

$$M_k = A_1^k M_0 \quad k > 0 \tag{4.38}$$

which gives

$$\hat{A}_1 = \hat{M}_1 \hat{M}_0^{-1} \quad k > 0 \tag{4.39}$$

B matrix is obtained by solving the following equation:

$$\hat{B}\hat{B}^T = \hat{M}_0 - \hat{A}_1 \hat{M}_1^T \tag{4.40}$$

The multivariate AR(2) model is presented as

$$Z_t = A_1 Z_{t-1} + A_2 Z_{t-2} + B\varepsilon_t \tag{4.41}$$

The parameters of a multivariate AR(2) model are obtained as follows:

$$\hat{A}_1 = \left[\hat{M}_1 - \hat{M}_2 \hat{M}_0^{-1} \hat{M}_1^T\right] \left[\hat{M}_0 - \hat{M}_1 \hat{M}_0^{-1} \hat{M}_1^T\right]^{-1} \tag{4.42}$$

$$\hat{A}_2 = \left[\hat{M}_2 - \hat{M}_1 \hat{M}_0^{-1} \hat{M}_1\right] \left[\hat{M}_0 - \hat{M}_1^T \hat{M}_0^{-1} \hat{M}_1\right]^{-1} \tag{4.43}$$

$$\hat{B}\hat{B}^T = \hat{M}_0 - \left[\hat{A}_1 \hat{M}_1^T + \hat{A}_2 \hat{M}_2^T\right] \tag{4.44}$$

4.3.8 Simulation and Validation

As it was described in the introduction of this chapter, two major applications are aimed by the time series modeling: forecasting and synthetic data generation. Forecasting is the process of making statements about events whose actual outcomes have not yet been observed. Synthetic data generation is any production of data applicable to a given situation that are not obtained by direct measurement. Usually forecast data are used for operation of water resources and environmental systems, where synthetic generated data are used for simulation of such systems for designing purposes.

There are two main differences between forecasting and data generation:

1. Data generation has a long-term time horizon (say 5 years to 100 years), whereas forecasting is usually applied for a short-term time horizon (say 1 month, 3 months, and maximum 1 year).
2. Forecast data refers to a specific time, but a generated data may refer to any time within a long time interval.

Even though there are models that outperform ARMA models in forecasting, ARMA is still one of the top choices of modelers for synthetic data generation.

4.3.8.1 Forecasting

The algorithmic procedure of forecasting by an ARMA or ARIMA model is presented in Fig. 4.14. As it is shown in the figure, the procedure contains six major steps which are presented as follows:

Fig. 4.14 The algorithm of forecasting by time series models

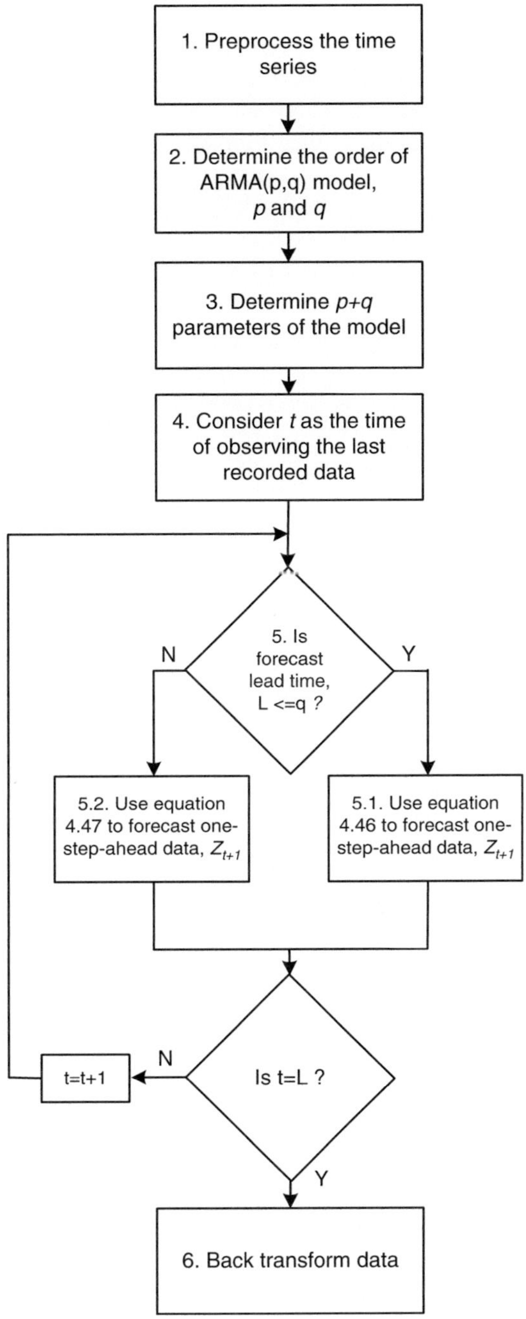

1. Preprocessing: Normalize and standardize n numbers of recorded data at the beginning of the procedure. In case of applying ARIMA model, the data should be differenced by a specific order. The order of differencing depends on the order that the data is nonstationary.
2. Order of the model: Plot ACF and PACF to decide on the appropriate range for the orders of the model.
3. Parameters of the model: Fit a model of ARMA(p,q) or ARIMA(p,d,q) to the above-transformed data and determine their parameters.
4. Setting up the forecasting model: The following general equation is set up based on the type and order of the model as well as the calculated parameters. The general relation for time series forecasting is

$$U_{t+1} = \sum_{i=1}^{p} \varphi_i U_t - \sum_{j=1}^{q} \theta_j \varepsilon_{t+1-j} \tag{4.45}$$

U in the above equation refers to either normal standard data or the differenced normal standard data. It is to be notified that t in the equation refers to the time of observing the last recorded data. The ε_t values prior to t are actually the errors of modeling in the previous steps. The random number of the ARMA equation has been omitted from the equation since the forecasting model relied only on the deterministic term of a time series.
5. Use the correct equation: Considering the lead time of forecasting as L, if $L \leq q$ the following equation is used for forecasting. It is because in case of no information about the future random variables, we use their expected value which is in fact zero.

$$U_{t+1} = \sum_{i=1}^{p} \varphi_i U_t - \sum_{j=1}^{q} \theta_j \varepsilon_{t+1-j} \tag{4.46}$$

If $L > q$ the forecasting equation changes to the following:

$$U_{t+1} = \sum_{i=1}^{p} \varphi_i U_t \tag{4.47}$$

6. Back-transform data: The forecasted data should be back-transformed to the real values.

For a time series forecasting by model m, the following syntax is used:

```
Zp = predict(m,Z,L);
```

where m is the selected model, Zp is the forecasted series, and L is the lead time of forecasting.

4.3.8.2 Synthetic Data Generation

The algorithmic procedure of synthetic data generation by an ARMA model is presented in Fig. 4.15. It is to be notified that an ARIMA model is not applicable to the data generation purposes, because the differencing step of ARIMA makes it necessary to deal with the exact time location of a data to be integrated with its previous variable. This is in contrast with the idea of data generation in which no specific time is considered for a generated data. The procedure of data generation is summarized as follows:

1. Preprocessing: At the beginning of the procedure, normalize and standardize n numbers of recorded data.
2. Order of the model: Plot ACF and PACF to decide on the appropriate range for the orders of the model.
3. Parameters of the model: Fit a model of ARMA(p,q) to the above-transformed data and determine their parameters.
4. Random number generation: If you wish to generate m numbers of synthetic data, generate 50–100 + m numbers of normal standard random variables with average zero and standard deviation equal to the standard deviation of the basic model errors. The first 50–100 random numbers are used to warm up the model and to reduce the effect of initial choices for the process of data generation.
5. Setting up the model: The following general equation is set up based on the type and order of the model as well as the calculated parameters.

$$Z_{t+1} = \sum_{i=1}^{p} \varphi_i Z_t - \sum_{j=1}^{q} \theta_j \varepsilon_{t+1-j} + \zeta_{t+1} \tag{4.48}$$

   It is to be notified that t in the equation refers to the time of observing the last recorded data.
6. Data generation: Calculate the first synthetic data by Eq. (4.48). The calculation of Z_{t+1} continues to the time horizon of 50–100 + m, where m = number of desired synthetic data. At each iteration, Z_t and ε_t are replaced recursively by Z_{t+1} and ε_{t+1} of the previous iteration.
7. Omitting the warm-up data: The first 50–100 generated data which were used as warm up are omitted in this stage.
8. Back-transform data: The generated data are back-transformed to the real values.

The role of lead time in time series forecasting and data generation is shown in Fig. 4.16. As it is shown in Fig. 4.16a, two different relations are used for time series forecasting before the lead time of $t + q$ and after that. The role of lead time in synthetic data generation is shown in Fig. 4.16b. As it is shown in the figure, considering the current time as t, data generated in the range of $50 + t$ to $100 + t$ is omitted from the final data and the set of data after that is considered as the final data.

Fig. 4.15 The algorithm of synthetic data generation by time series models

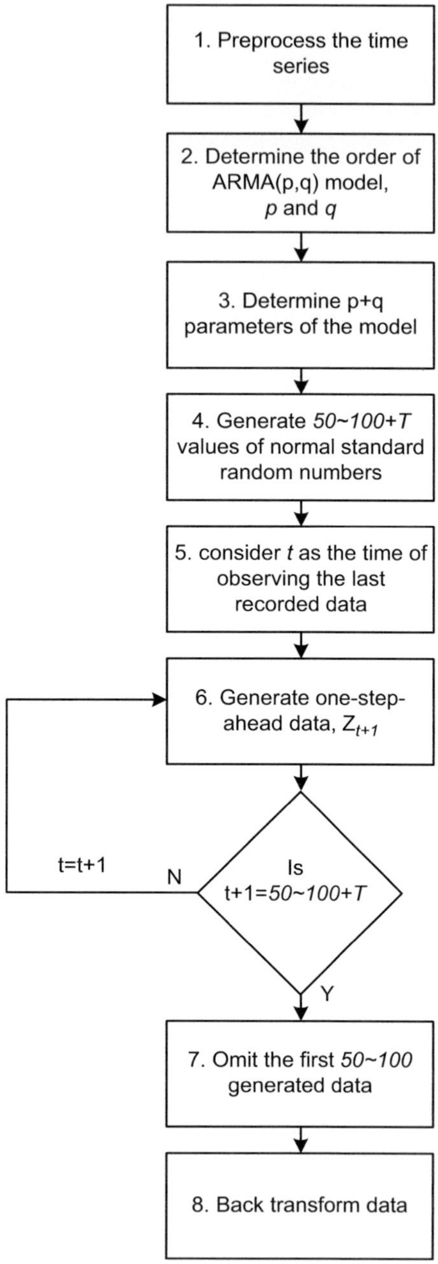

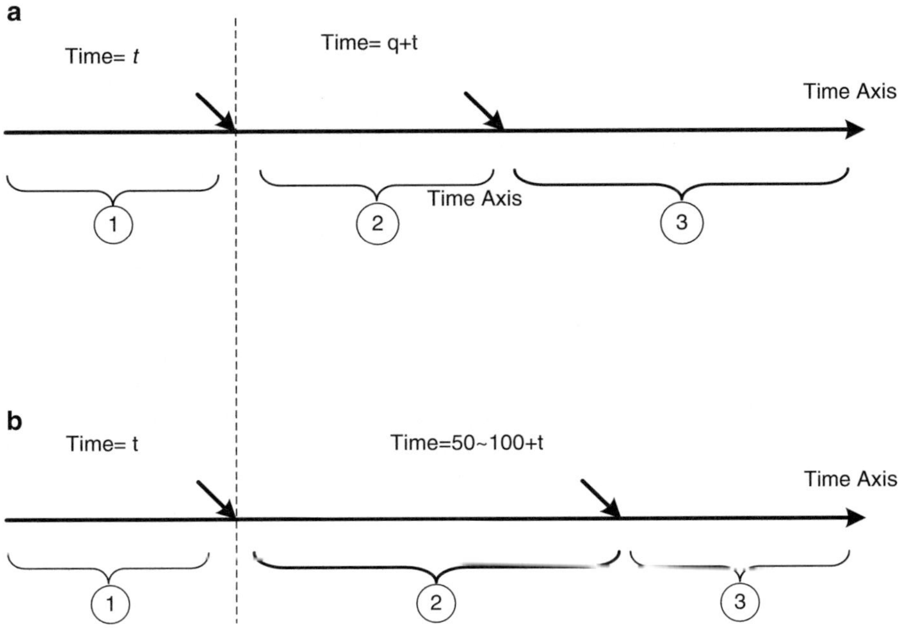

Fig. 4.16 The role of time intervals in time series forecasting and data generation (**a**) Forecasting (**b**) Data generation

4.3.8.3 Validation of the Models

The validation of ARMAX models could be tested by the Akaike Information Criterion (AIC) which has been suggested by Akaike (1974) using the following relation:

$$AIC = N \ln \sigma_{\varepsilon}^{2} + 2(p + q) \tag{4.49}$$

where N = number of time series data, σ_{ε}^{2} = variance of the error of the model, and p and q = the orders of the model. AIC considers two criteria to decide on a validation of a model, the minimum error (first sentence of Eq. (4.49)), and parsimony of the parameters (second sentence of Eq. (4.49)). Obviously the model with less AIC is preferred. To obtain the value of this test for a given model, m, the following syntax is used:

```
test=aic(m);
```

4.4 Summary

Modeling a stochastic time series, $X_t = P_t + Tr_t + Z_t$, includes dealing with at least three terms of P_t as a deterministic periodic term, Tr_t as a deterministic linear or nonlinear trend, and Z_t as a stochastic term. Preprocessing of a time series involves exploring periodicity, trend, and correlation between the data of a time series. Furthermore, a time series might have a sudden significant change in its long-term average which is known as "*jump*." A time series should be stationary type I by removing jump as well as P_t and Tr_t before being processed through the component of Z_t. Two tests were presented for a time series, namely, *Mann–Kendall* test of trend and *Kruskal and Wallis* test of jump.

Two major applications are aimed by the time series modeling: forecasting and synthetic data generation. Forecasting is the process of making statements about events whose actual outcomes have not yet been observed. Synthetic data generation is any production of data applicable to a given situation that are not obtained by direct measurement. Usually forecasted data are used for operation of water resources and environmental systems, where synthetic generated data are used for simulation of such systems for the designing purposes.

Time series analysis accounts for the fact that data points taken over time may have an internal structure (such as autocorrelation, trend, or seasonal variation) that should be dealt with. The process of time series modeling consists of four major steps:

- Model selection among the range of ARMA (p,q), ARIMA(p,d,q), and ARMAX(p,q,b) models
- Selection of the order of models among the autocorrelation, partial autocorrelation, and cross-correlation orders
- Determination of the parameters of the models such as φ and θ parameters
- Simulation and validation for data generation and forecasting

The basis for autoregressive, AR(p), models is the Markov chain, which relates to the time series where every individual data is corrected with its past and future data. A Markov chain is a mathematical system that undergoes transitions from one state to another, between a finite and countable number of possible states. The experience of applying AR(p) models within the real-world data might result in significant errors. The reason for the magnitude of the unacceptable error is related to the fact that the occurrence of Z_t in the current time is considered only as a function of its previous values. It is a fact that the effect of other predictors is neglected in the conventional AR(p). A solution to mitigate this shortcoming of the *AR* models is to consider the error term of the model, ε_t, as a representative of all other predictors except the persistency of the time series data. This idea improves an AR model to an ARMA model. An autoregressive integrated moving average (ARIMA) model is a generalization of an autoregressive moving average (ARMA) model. They are applied in cases where data show evidence of non-stationarity, where an initial differencing step can be applied to remove the non-stationarity. To consider the effect of an external predictor except the persistence between data, an improved model of ARMA, ARMAX (AutoRegressive Moving Average with eXogenous inputs) has been developed. In general, the *ARMAX*(p,q,b) model might be expected to simulate very well the behavior of systems whose input–output characteristics are approximately linear. Multivariate time series analysis is used when one wants to model and explain the interactions and co-movements among a group of time series variables. In many problems of water resources and environmental engineering, it is essential to model

Table 4.13 Summary of the classic time series models

Model	Basic correlation analysis	Applicable areas
Autoregressive AR(p) model	Partial autocorrelation function	Synthetic data generation, forecasting
Autoregressive moving average ARMA(p,q) model	Autocorrelation function and partial autocorrelation function	Synthetic data generation, forecasting
Autoregressive integrated moving average ARIMA(p,d,q) model	Autocorrelation function and partial autocorrelation function	Forecasting of time series with slight trend
Autoregressive–moving average model with exogenous inputs ARMAX(p,q,b)	Cross-correlation function, autocorrelation function, and partial autocorrelation function	Synthetic data generation and forecasting getting benefit of time series of an additional predictor
Multivariate autoregressive, MAR model	Cross-correlation function and partial autocorrelation function	Synthetic data generation and forecasting

two or more time series simultaneously due to the correlation between multiple variables. The validation of the group of ARMA and ARMAX models can be tested by the *Akaike* Information Criterion (AIC). Table 4.13 demonstrates a summary of the characteristics and applicable areas of the models.

It is to be notified that time series modeling is not limited to the models presented in this chapter. Actually, regression-based models as presented in Chap. 3, especially K-NN model, as well as dynamic artificial neural networks as will be presented in Chap. 6 also could be used for this purpose. What were presented in this chapter were actually the basic stochastic time series models used especially for synthetic data generation.

Workshop

Figure 4.17 shows the schematic of a two-reservoir system of surface water resources which supplies potable water needs as well as agricultural demands. For the purpose of modeling the reliability of the system, we need to extend 50 years annual data of inflow to the reservoirs. Furthermore, for the operation of the system, a model is needed to forecast 1-year ahead data. This workshop reviews the application of time series techniques to deal with the explained purposes (Table 4.14).

Step 0. Preprocessing of the Data

First, we investigate trend and jump in the time series of rivers 1 and 2. The Mann–Kendall and Kruskal–Wallis tests demonstrate that no trend and jump exist in both time series. Also, since the time series is not a seasonal time series, no periodicity is considered for the time series (Fig. 4.18).

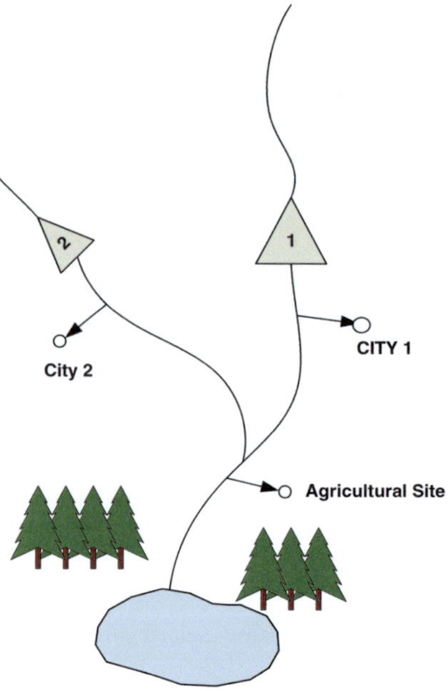

Fig. 4.17 Schematic of the study area of the workshop

Table 4.14 Time series presented for the workshop

Year	1	2	Year	1	2
1	5,605.6	1,765.0	26	12,231.8	2,834.6
2	7,865.0	3,212.0	27	4,967.2	2,492.6
3	8,765.0	2,341.0	28	9,876.0	2,972.4
4	9,876.0	2,141.5	29	6,271.2	2,946.2
5	10,234.0	2,341.0	30	10,739.6	3,106.6
6	8,976.0	2,280.0	31	12,768.0	4,994.0
7	8,965.0	2,946.2	32	5,761.0	3,128.0
8	6,727.7	1,722.3	33	6,075.5	2,929.2
9	9,876.0	2,345.0	34	5,509.8	1,752.3
10	8,296.4	1,316.9	35	15,160.5	5,964.4
11	9,136.0	2,679.5	36	7,221.5	2,946.2
12	10,234.0	3,249.0	37	16,461.2	3,476.3
13	11,143.0	3,208.7	38	15,033.8	5,291.3
14	7,167.4	3,014.3	39	8,112.1	4,023.6
15	14,969.9	5,681.5	40	7,415.2	2,854.0
16	6,145.0	3,473.5	41	13,888.0	3,983.3
17	14,230.0	5,463.0	42	6,780.0	1,757.0
18	8,018.1	3,368.5	43	6,578.0	1,103.5
19	12,302.7	5,425.2	44	6,547.0	847.5
20	4,893.1	2,511.8	45	7,860.0	3,212.0
21	9,543.1	2,770.8	46	9,240.6	2,236.6
22	4,734.6	2,137.1	47	8,254.0	2,341.8
23	11,770.8	5,226.6	48	8,970.0	2,112.6
24	10,953.7	3,573.2	49	6,720.0	2,871.6
25	9,358.8	2,888.5	50	7,890.0	2,543.9

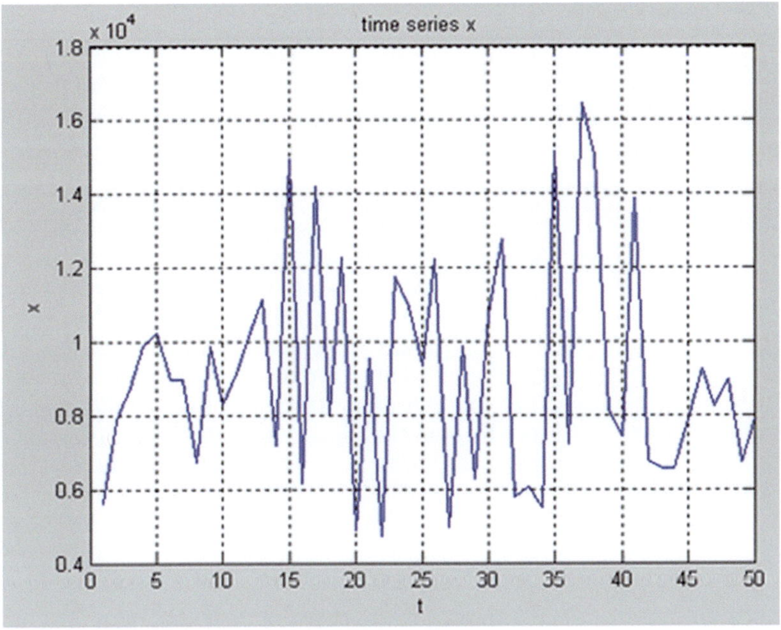

Time series 1

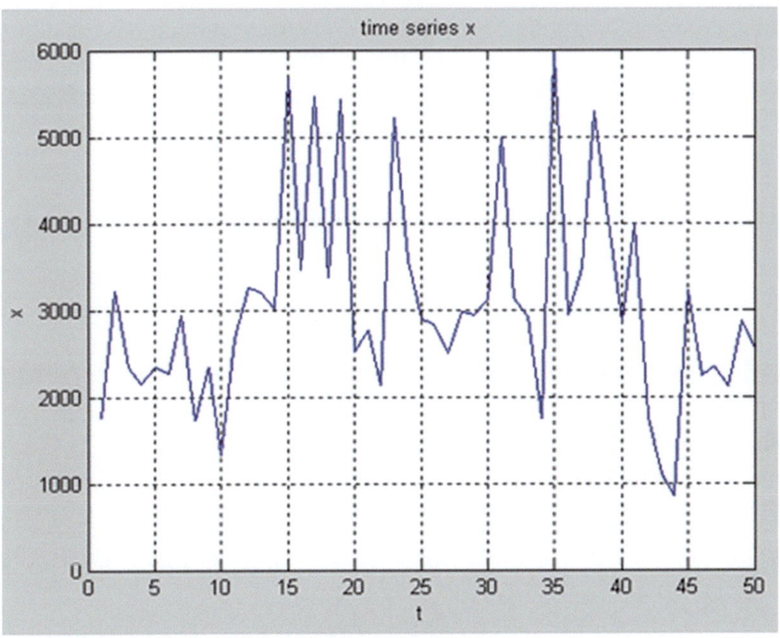

Time series 2

Fig. 4.18 Time series presented for the workshop

Table 4.15 Normal standard time series of the workshop

Year	Z1	Z2	Year	Z1	Z2
1	−1.3946	−1.1608	26	1.0752	−0.0445
2	−0.2900	0.2782	27	−1.8017	−0.3637
3	0.0527	−0.5150	28	0.4243	0.0766
4	0.4243	−0.7247	29	−1.0229	0.0539
5	0.5340	−0.5150	30	0.6816	0.1908
6	0.1273	−0.5777	31	1.2035	1.5219
7	0.1234	0.0539	32	−1.3035	0.2087
8	−0.7930	−1.2142	33	−1.1274	0.0391
9	0.4243	−0.5109	34	−1.4523	−1.1766
10	−0.1204	−1.7738	35	1.7094	2.0719
11	0.1825	−0.1857	36	−0.5636	0.0539
12	0.5340	0.3083	37	1.9476	0.4888
13	0.7939	0.2755	38	1.6849	1.6977
14	−0.5878	0.1126	39	−0.1916	0.8919
15	1.6725	1.9184	40	−0.4784	−0.0272
16	−1.0898	0.4866	41	1.4526	0.8636
17	1.5242	1.7962	42	−0.7678	−1.1707
18	−0.2286	0.4043	43	−0.8664	−2.1180
19	1.0925	1.7747	44	−0.8819	−2.5983
20	−1.8528	−0.3450	45	−0.2920	0.2782
21	0.3181	−0.1019	46	0.2180	−0.6231
22	−1.9651	−0.7295	47	−0.1366	−0.5141
23	0.9597	1.6601	48	0.1252	−0.7562
24	0.7418	0.5632	49	−0.7968	−0.0116
25	0.2575	0.0033	50	−0.2799	−0.3140

Each time series is normalized using Box–Cox transformation as follows:

```
[Y, lambda] = boxcox(data)
```

The results of *lambda* for each river are obtained as

Lambda for river 1 = −0.1379
Lambda for river 2 = 0.3074
The time series are standardized then by the following syntax:

```
Z=(Y-mean(Y))/std(Y);
```

The transformed time series which are now normal standard are presented in Table 4.15. The time series are now ready to be used through the modeling process.

Step 1. Determining the Order of Time Series Models

(A) *Univariate Approach*

Autocorrelation function (ACF) and partial autocorrelation function (PACF) by lag 15 are plotted for Z1 and Z2 normal standard time series using the following syntaxes:

```
Lag=15;
autocorr(Z1,Lag)
autocorr(Z2,Lag)
parcorr(Z1,Lag)
parcorr(Z2,Lag)
```

The results are presented in (Figs. 4.19 and 4.20):

The plots demonstrate that the range of acceptable orders of an ARMA model for Z1 is ARMA(1,3) and for Z2 time series is ARMA(2,2).

(B) *Multivariate Approach*

Cross-correlation between two time series is calculated by the following syntax for lag time 15, which results in the following function:

```
crosscorr(Z1,Z2,15)
```

The result shown in Fig. 4.21 demonstrates that the data of two stations are significantly correlated in lag 0. It means that using a multivariate model is preferred in this problem.

Step 2. Determining the Parameters of the Models

The following program finds the best orders of an ARMA(p,q) model for a time series based on the AIC. Please note that the following program searches the best model among the range of ARMA(2,2) which includes AR (1), MA(1), ARMA(1,1), ARMA (2,1), ARMA(1,2), and ARMA(2,2). The program could be easily extended to higher orders of models by changing *pvec* and *qvec*.

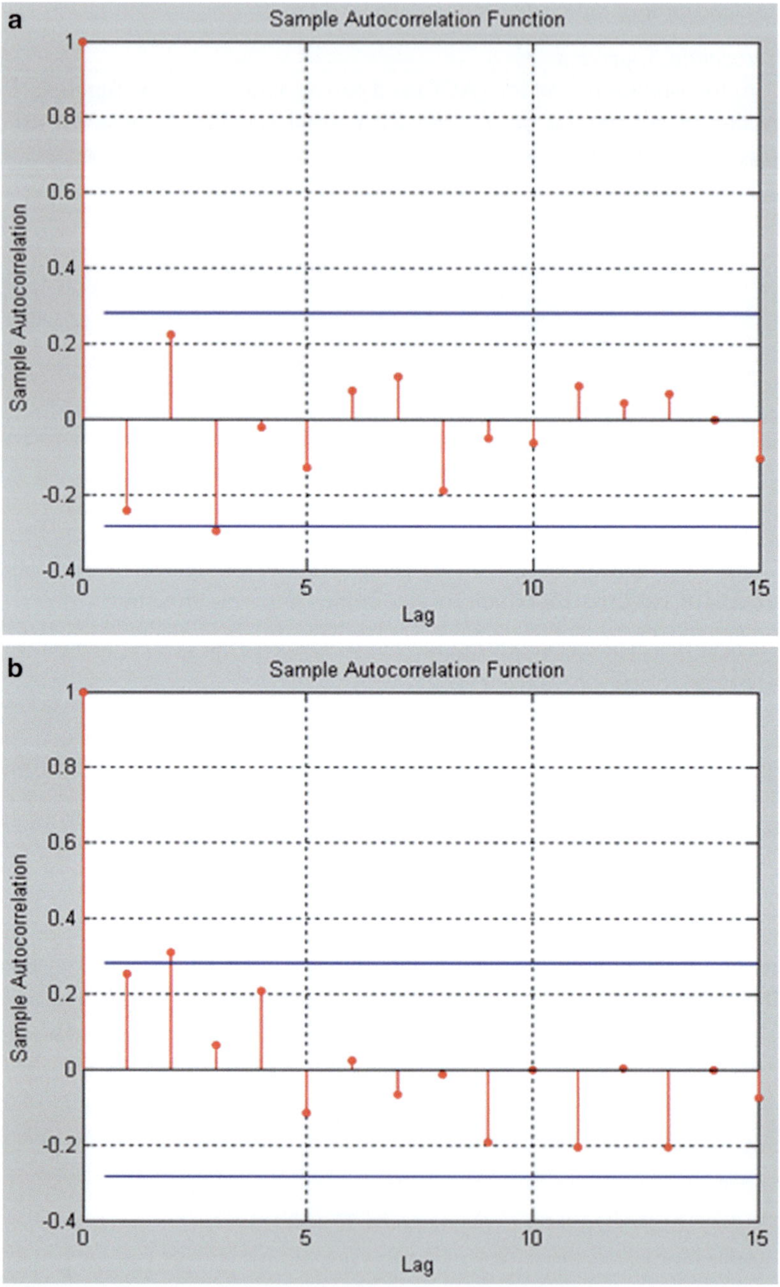

Fig. 4.19 Autocorrelation function for time series Z1 (**a**) and Z2 (**b**)

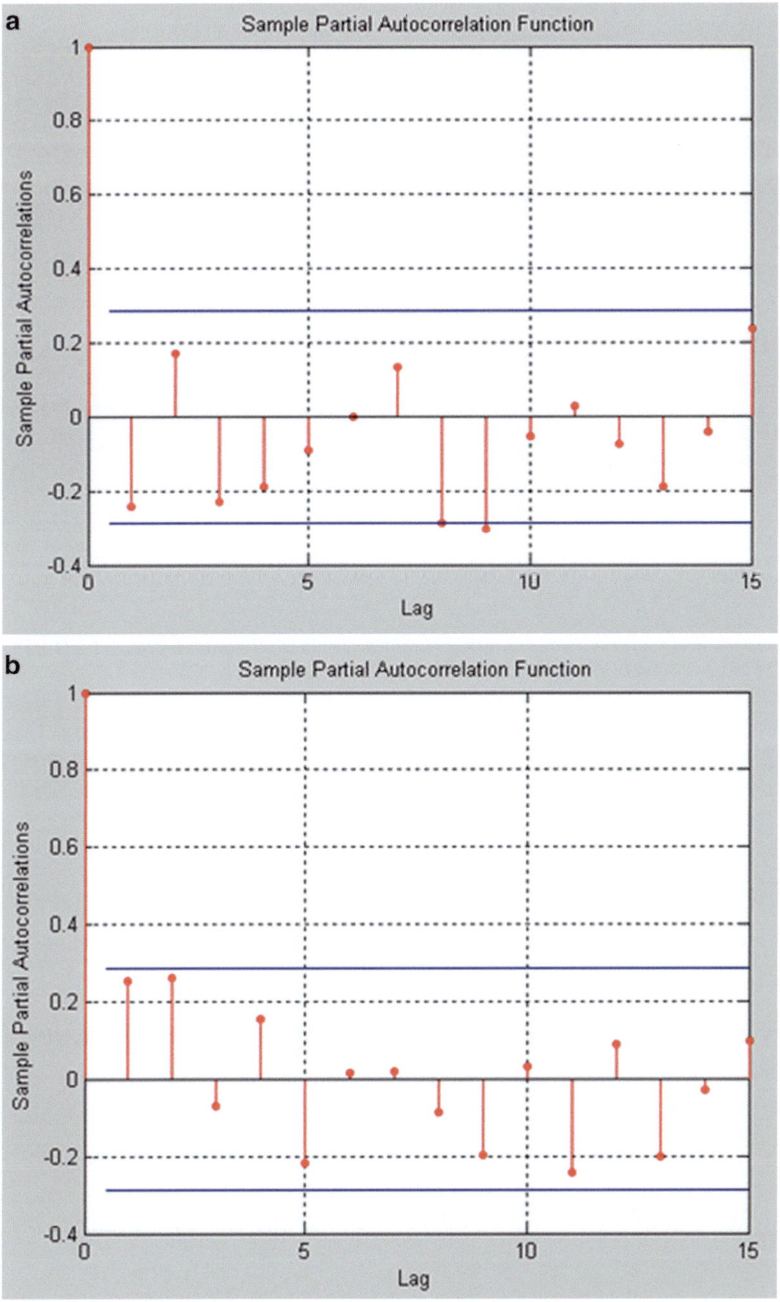

Fig. 4.20 Partial autocorrelation function for time series Z1 (**a**) and Z2 (**b**)

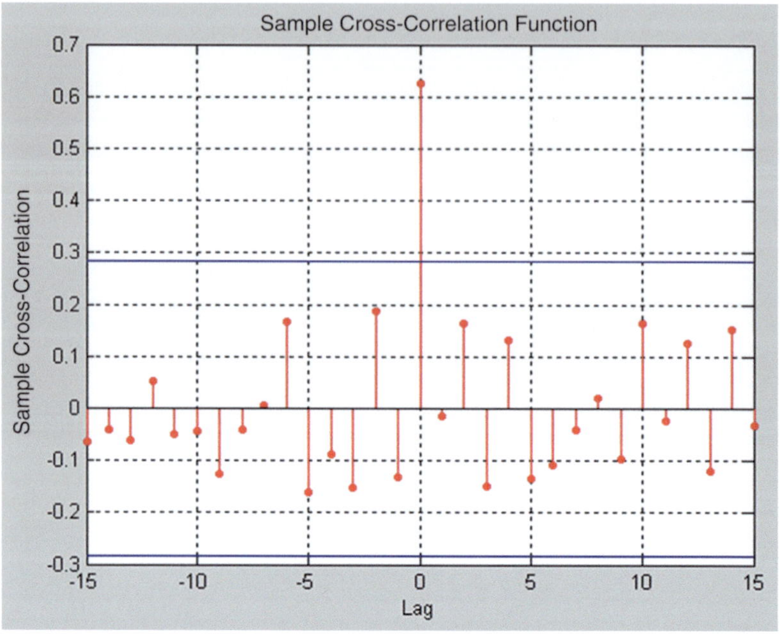

Fig. 4.21 Cross-correlation function for time series Z1 and Z2

```
%%program to determine the orders of an ARMA model >>

pvec=[0,1,2]; % orders of AR
qvec=[0,1,2]; % orders of MA

%% pvec and qvec are the orders of AR and MA,
%% respectively, which are going to be tested trough
%% the program


%% parameters and variables which are used in the
%% program are presented here

np=length(pvec);
nq=length(qvec);
aicsave=-99*ones(np,nq);
fpesave=-99*ones(np,nq);
minaic=1e+6;
```

(continued)

```
%% loop to test different models
%% using Akaike Information Criteria (AIC)

for pp=1:np
    p=pvec(pp);
    for qq=1:nq
        q=qvec(qq);
        if p+q ~=0
           orders=[p q];
            m=armax(Z,orders);

% m is a structure, which contains the parameters
% associated with a specific order

      aicsave(pp,qq)=aic(m);

%% aicsave, saves the AIC associated with the
%% model, m.

      fpesave(pp,qq)=fpe(m);
       if aicsave(pp,qq) < minaic
          minaic=aicsave(pp,qq); % save the min
          pbest=p;
          qbest=q;
          mbest=m;

%% finally, mbest saves the structure of the
%% model with minimum AIC among the others

       end
     end
   end
end
```

Table 4.16 demonstrates variables and parameters which could be changed due to the change of the problem.

Table 4.16 Parameters and variable used in the program of the workshop

Variable/parameter	Description
Pvec	Order of the AR model
qvect	Order of the MA model
Z1	Name of the time series

Using the above program, the following results are obtained for each time series:

	Time series	
	Z1	Z2
Model	ARMA(1,1)	ARMA(2,1)
φ parameters	−0.7523	−0.581 and 0.4148
θ parameters	−0.5352	−0.9204

Step 3. Time Series Forecasting

Now we can forecast the streamflow of rivers *L*-step ahead by the following syntax:

```
Zp = predict(m,Z,L);
```

It should be notified that *m* in the above command refers to the best time series model fitted to the time series. Using the command, one can have the estimation of forecast variables only for the length of the given vector, *Z*. To forecast data in the lead time beyond the time of last recorded data, one can add L values at the end of vector *Z*. The values might be considered all equal to the average of the data. As another alternative Eqs. (4.46) and (4.47) could be used for forecasting.

Step 4. Synthetic Data Generation

Equation (4.48) could be used through the described process to generate synthetic data. As the process needs normal random numbers, the following syntax could be used to generate *r*q* matrix of random numbers if *mu* and *sigma* are considered as 0 and 1, respectively.

```
R = normrnd(mu,sigma,r,q)
```

References

Abaurrea J, Asin J, Cebrian AC, Garcia-Vera MA (2011) Trend analysis of water quality series based on regression models with correlated errors. J Hydrol 400:341–352

Akaike H (1974) A new look at the statistical model identification. IEEE Trans Autom Control 19 (6):716–723

Bayazit M, Önöz B (2007) To prewhiten or not to prewhiten in trend analysis? Hydrol Sci J 52 (4):611–624

Box GEP, Cox DR (1964) An analysis of transformations. J R Stat Soc Ser B 26(2):211–252

Burlando P, Rosso R, Cadavid LG, Salas JD (1993) Forecasting of short-term rainfall using ARMA models. J Hydrol 144:193–211

Cong Z, Yang D, Gao B, Yang H, Hu H (2009) Hydrological trend analysis in the Yellow River basin using a distributed hydrological model. Water Resour Res 45, W00A13 (13)

Darken PF, Zipper CE, Holtzman GI, Smith EP (2002) Serial correlation in water quality variables: estimation and implications for trend analysis. Water Resour Res 38:1117–1123

Hamed KH (2009a) Enhancing the effectiveness of prewhitening in trend analysis of hydrologic data. J Hydrol 368:143–155

Hamed KH (2009b) Exact distribution of the Mann–Kendall trend test statistic for persistent data. J Hydrol 365:86–94

Kendall MG (1975) Rank correlation methods. Charles Griffin, London

Kruskal WH, Wallis WA (1952) Use of ranks on one-criterion variance analysis. J Am Stat Assoc 47:583–621 (correction appear in vol 48, pp 907–911)

Landeras G, Ortiz-Barredo A, López JJ (2009) Forecasting weekly evapotranspiration with ARIMA and artificial neural network models. J Irrig Drain Eng 135(3):323–334

Mann HB (1945) Nonparametric tests against trend. Econometrica 13:245–259

Mohammadi K, Eslami HR, Kahawita R (2006) Parameter estimation of an ARMA model for river flow forecasting using goal programming. J Hydrol 331:293–299

Salas JD, Delleur JW, Yevjevich V, Lane WL (1980) Analysis and modeling of hydrologic time series. In: Maidment DR (ed) Hand book of hydrology. McGrow-Hill, New York

Şen Z (2011) An innovative trend analysis methodology. J Hydrol Eng 17(9):1042–1046

Shao Q, Li M (2011) A new trend analysis for seasonal time series with consideration of data dependence. J Hydrol 396:104–112

Song AN, Chandramouli V, Gupta N (2011) Analyzing Indiana Reservoirs inflow trend using self-organizing map. J Hydrol Eng 17(8):880–887

Sun H, Koch M (2001) Case study: analysis and forecasting of salinity in Apalachicola Bay, Florida, using Box-Jenkins ARIMA models. J Hydraul Eng 127(9):718–727

Walker G (1931) On periodicity in series of related terms. Proc R Soc Lond Ser A 131:518–532

Yue S, Pilon P (2004) A comparison of the power of the t test, Mann-Kendall and bootstrap tests for trend detection. Hydrol Sci J 49(1):21–37

Yule GU (1927) On a method of investigating periodicities in disturbed series, with special reference to Wolfer's sunspot numbers. Philos Trans R Soc Lond Ser A 226:267–298

Chapter 5
Artificial Neural Networks

Abstract Artificial neural network as the most famous artificial intelligence models are a collection of neurons with specific architecture formed based on the relationship between neurons in different layers. Neuron is a mathematical unit, and an artificial neural network that consists of neurons is a complex and nonlinear system. Artificial neural networks (ANNs) may have different architectures which result in different types of ANNs. A static ANN known as a multilayer perceptron (MLP) is the most applied ANN in different fields of engineering. This type of ANN is presented in this chapter and details on its calibration and validation are discussed. Furthermore, dynamic ANNs improved to consider the temporal dimension of data through the modeling process is presented. In this chapter, dynamic ANNs including input delay networks, recurrent networks, and a combination of both are discussed in details. Statistical neural networks, namely, radial basis estimator, generalized neural network, and probabilistic neural network, which are all developed based on a statistical-based estimation, are the third type of ANNs presented in this chapter. How to deal with calibration and validation of all models by MATLAB codes and commands are discussed. Application of the models in function approximation and data classification are presented through different examples.

Keywords Artificial neural networks • Dynamics neural networks • Statistical neural networks • Radial basis function • Probabilistic neural network

5.1 Introduction

Artificial neural networks (ANNs) are models based on the structure of the human brain and are used for complicated problems of pattern recognition, clustering, classification, and simulation. It has been proven that ANNs are universal function approximators that are capable of mapping any complicated nonlinear function. ANNs are able to intelligently learn those functions through a training process. The capability of ANNs for mapping a set of input/output data with an

S. Araghinejad, *Data-Driven Modeling: Using MATLAB® in Water Resources and Environmental Engineering*, Water Science and Technology Library 67,
DOI 10.1007/978-94-007-7506-0_5,

Table 5.1 A summary review on the application of ANNs in water resources and environmental engineering

Field of the study	Researchers
Evapotranspiration modeling	Trajkovic et al. (2003), Kisi and Yıldırım (2007)
Flood forecasting	Toth et al. (2000)
Hydrological prediction	Thirumalaiah and Deo (2000), Anmala et al. (2000)
Impacts of climate change on water supply	Elgaali and Garcia (2007)
Rainfall–runoff modeling	Garbrecht (2006)
Surface reservoir modeling	Chandramouli and Raman (2001), Neelakantan and Pundarikanthan (2000)
Water quality modeling	Schmid and Koskiaho (2006), Suen and Eheart (2003), Milot et al. (2002)

acceptable range of error makes them useful tools for natural processes modeling. In the recent years, artificial neural networks have been widely studied and applied in the field of water resources and environmental engineering. Similar to most of the data-driven models, ANNs do not generate parametric relationship between the independent and dependent variables. Instead, they provide a relationship between input and output data through a training process to approximate any continuous functions with a specific accuracy. The ANN approach is an effective and efficient way to model data-dependent problems in situations where the explicit knowledge of the internal physical subprocess either is not required or it is not still discovered.

Application of the artificial neural networks in the field of water resources and environmental engineering has grown fast in the recent decade. The number of articles published in the subject of ANNs in water resources and environmental engineering has a significant increasing trend in the last two decades. ANN has been of interest in the contests such as rainfall–runoff modeling, streamflow forecasting, and groundwater modeling. Table 5.1 shows a summary review on the fields that various researchers have found the application of ANNs useful.

This chapter is structured as shown in the chart of Fig. 5.1. After the "*Introduction*," basic definition is presented containing "*Components of an ANN*," "*Training Algorithm*," and "*Mapping by ANNs*." Obviously, the section put stress on the "*mapping functions*" capability of ANNs. The next section deals with the introduction of famous ANN models, which are widely used in different subjects. Theoretical background, networks architecture, their training and simulation methods, as well as the codes necessary for applying the networks are presented in this section. After presenting each network sample, applications are discussed in different illustrative examples. Static and dynamic networks, as well as statistical networks, are those which are described in that section. Finally, the chapter ends with a workshop.

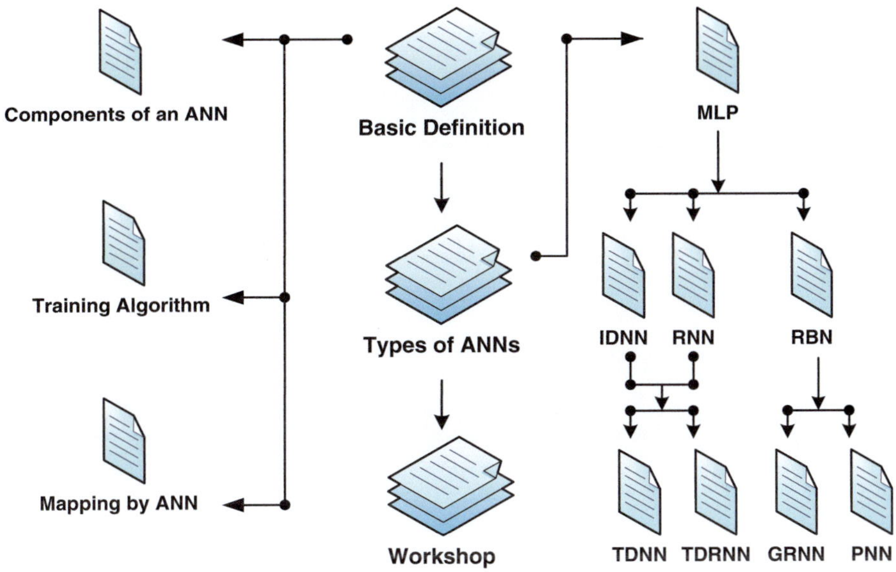

Fig. 5.1 The structure of this chapter

5.2 Basic Definitions

5.2.1 Components of an ANN

5.2.1.1 Neuron

The first big step toward neural networks was made in 1943 by McCulloch and Pitts in an article entitled "A logical calculus of the ideas immanent in nervous activity" (Anderson and Rosenfeld 1988). They were the first to present a mathematical model of the biological neuron as the basic switching element of the brain. This article laid the foundation for the construction of artificial neural networks. The "artificial neuron" is the basic unit of an artificial neural network, which is in turn a surrogate of the biological neuron. It is necessary to understand the computational capabilities of this processing unit as a prerequisite for understanding the function of a network of such units. A biological neuron consists of four major parts, namely, *soma*, *dendrite*, *axon*, and *synapse* as shown in Fig. 5.2.

The cell body of the neuron (*soma*) can store small electrical charges, similarly to a battery. This storage is loaded by incoming electrical impulses from other neurons and through *dendrites* (Ertel and Black 2011). The more electric impulse comes in, the higher the voltage. If the voltage exceeds a certain threshold, the neuron will fire. This means that it unloads its store, in that it sends a spike over the axon and the synapses. The electrical current divides and reaches many other neurons over the synapses, which magnify the current and the same process takes place. The idea of developing an artificial neuron uses the same process by considering neuron as a unit that acts two

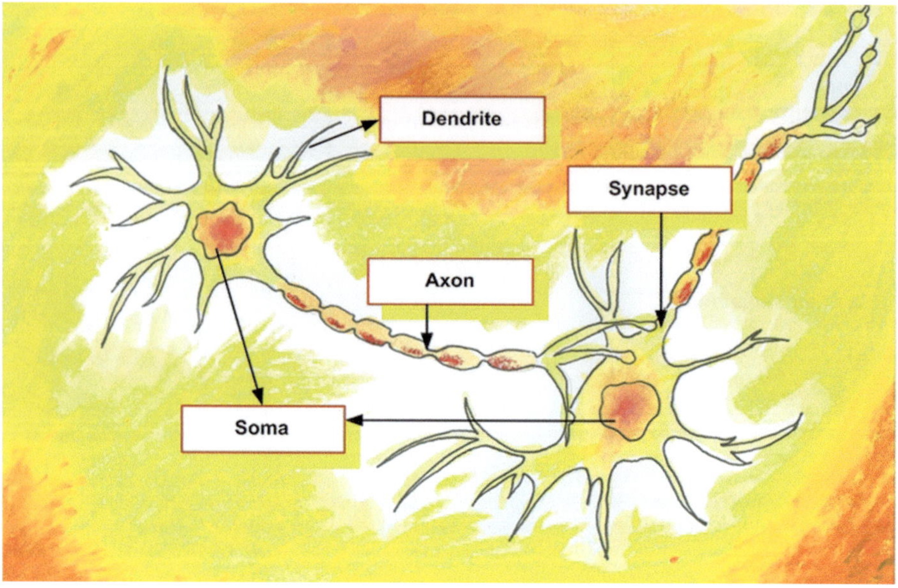

Fig. 5.2 Schematic of a biological neuron

Fig. 5.3 Schematic of an artificial neuron

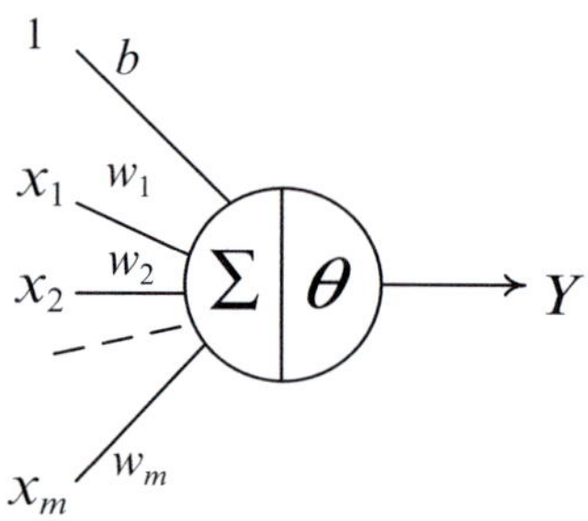

roles of combining the inputs that come in (X) and comparing the combined inputs with a specific threshold (θ) to determine appropriate output (Fig. 5.3).

Like synapses that control the magnitude of each single input, the inputs to an artificial neuron could be weighted by a weight matrix. For a mathematical reason, which is described in the next section, an artificial neuron usually gets benefit from an additional unit input with a weight known as *bias*. A comparison between the components of a neural cell and an artificial neuron is demonstrated in Table 5.2.

The mathematical relation of the functional process of an artificial neuron is defined as

$$\begin{aligned} I &= W \times X + b \\ Y &= \begin{cases} 1 & \text{if } I \geq \theta \\ 0 & \text{if } I < \theta \end{cases} \end{aligned} \tag{5.1}$$

where X = inputs, W = weight matrix, b = bias, I = sum of the weighted inputs, θ = threshold, and finally Y = output. The whole processing unit described above is called *perceptron*.

Table 5.2 Comparison between the components of a biological neuron and an artificial neuron

Neural cell	Artificial neuron
Soma	Neuron
Dendrite	Input
Synapse	Weights
Axon	Output

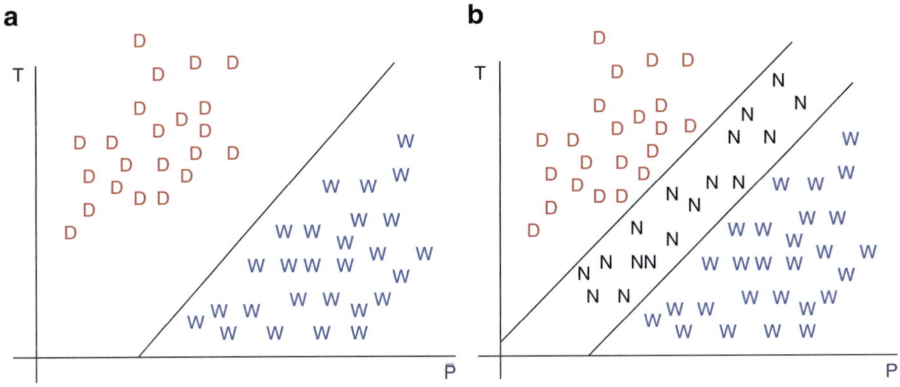

Fig. 5.4 Historical values of dry, wet, and normal years (shown by *D*, *W*, and *N*, respectively) associated with averaged precipitation and air temperature (**a**) Dry and wet years (**b**) Dry, wet, and normal years

5.2.1.2 Multilayer Perceptron

A single neuron is insufficient for solving many practical problems, mathematically. So a network of perceptrons is frequently used in parallel and series which is called neural network. The following example shows how the number of neurons could improve the ability of a network in solving practical problems.

Example 5.1: Clustering by ANNs

To cluster the climatic situation of the coming season of a basin as a wet or dry condition, the predicted values of averaged precipitation (P) and air temperature (T) are used. Several observed historical values of dry and wet years according to the associated Ps and Ts are drawn in Fig. 5.4a. How could a neuron be used for this clustering problem?

Solution

As it is demonstrated by Fig. 5.4a, dry and wet years are separable by a line. Suppose that the equation of the line is

$$w_1P + w_2T + b = 0$$

Those (P, T) points, which lay in the left side of the line, result in $w_1P + w_2T + b > 0$ and represent a dry cluster. Those points which lay in the right side of the line result in $w_1P + w_2T + b < 0$ and represent a wet season. Let us consider 1 and 0 as representatives of dry and wet years, respectively. So, the neuron that fits this problem is considered aswhere the mathematical expression of the neuron is

$$Y = \begin{cases} \text{dry} = 1 & \text{if } I \geq 0 \\ \text{wet} = 0 & \text{if } I < 0 \end{cases}$$

where $I = w_1P + w_2T + b$

The perceptron is now ready to cluster each new values of (P, T) as dry or wet clusters.

Now, it is desired to extend the system to be able to cluster input data to three clusters of dry, wet, and normal seasons (Fig. 5.4b). Obviously it is not possible to divide the decision space to three clusters by a single line. However, it would be easy to separate those three clusters by the use of two different lines as shown in the figure. It is understood by the part one of this example that each neuron represents a specific line in two-dimensional decision space. Therefore, two neurons are needed to divide the decision space of Fig. 5.4b appropriately. Let us use 0, 1, and 2 for wet, normal, and dry clusters, respectively. The perceptron will then be (Fig. 5.6) where each neuron in the first layer represents a drawn line in Fig. 5.4b. In case that a (P, T) point lies in the dry zone, the output of both neurons will be 1. The final neuron collects the output of these neurons as $I = f(I)$, which results in 2, the representative of the dry cluster. When (P, T) lies in the wet zone, the output of both neurons is 0 and the final neuron reports 0 as the representative of a wet cluster. In case that a point lies in the normal zone, the output of one neuron is 0 where the output of the other neuron is equal to 1. The final report would be 1 which is the representative of the normal cluster.

It may be concluded that each neuron in the hidden layer represents a line, a page, or a hyper page in the decision space, regarding to the dimension of the input vector. Obviously, for a more complex problem, which needs a nonlinear mapping, more neurons are needed to set up the appropriate ANN.

The way neurons are connected determines how computations proceed and constitutes an important early design decision by a neural network developer. The most famous neural network is one that includes layers of parallel perceptrons which is known as *multilayer perceptron* or feedforward network. *Feedforward network* is a subclass of networks in which a connection is allowed from a node in a layer only to nodes in the next layer, as shown in Fig. 5.5. These networks are succinctly described by a sequence of numbers indicating the number of nodes in each layer. The layers between input and output layers are called hidden layers. For instance, the network shown in Fig. 5.7 is a 3-2-1

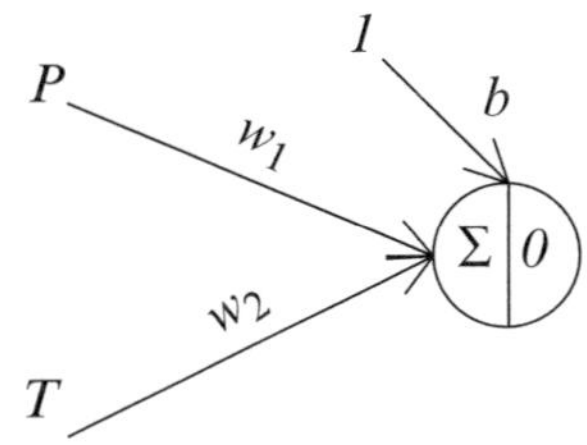

Fig. 5.5 Proposed neuron for part one of Example 5.1

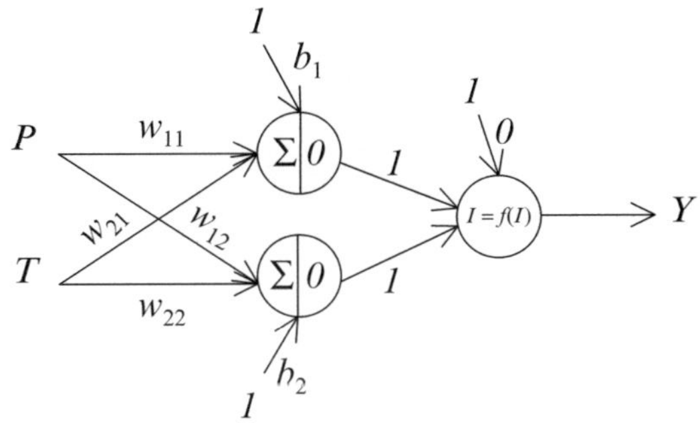

Fig. 5.6 Proposed neuron for part two of Example 5.1

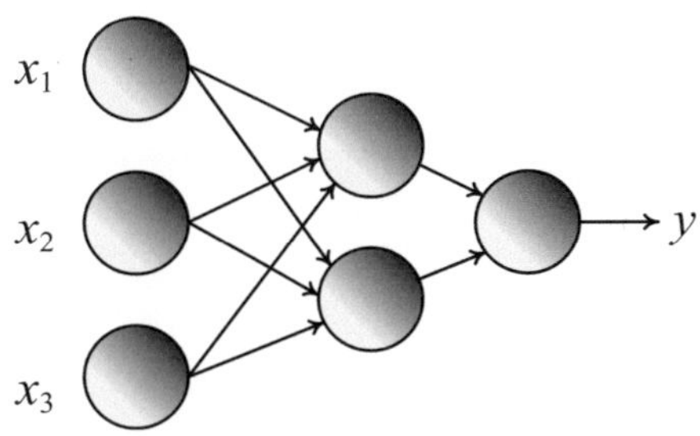

Fig. 5.7 A three-layer 3-2-1 feedforward network also known as a multilayer perceptron

feedforward network; it contains three nodes in the input layer (layer 0), two nodes in the first hidden layer (layer 1), and one node in the output layer (layer 2).

It should be noted that the three input nodes are not actually considered as neurons because they do not play a functional role.

5.2.1.3 Transfer Functions

As demonstrated in Example 5.1, the threshold value in the second half of a neuron could be replaced by a mathematical function to generalize the range of outputs that a neuron could produce. Actually, for computational purposes the function of a neuron could be generalized to any mathematical function that is called transfer function. A transfer function relates a particular input to an output as shown in the following figure (Fig. 5.8).

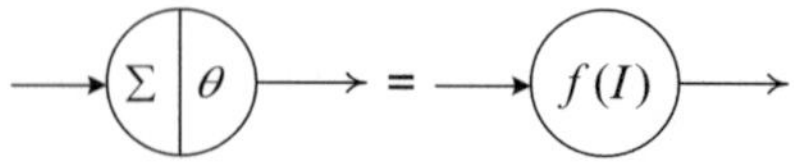

Fig. 5.8 Generalizing the function of a neuron

Table 5.3 Examples of transfer functions

Name	Function	Graphs
Linear	$f(x) = x$	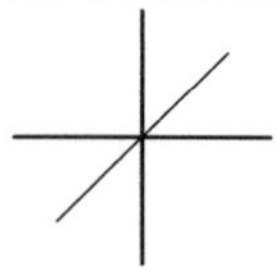
Symmetric–saturating–linear	$f(x) = \begin{cases} \delta & x \geq \theta \\ x & -\theta \leq x \leq \theta \\ -\delta & x \leq -\theta \end{cases}$	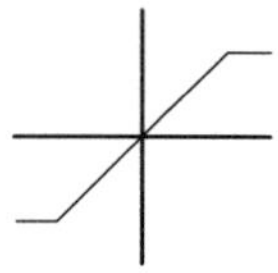
Log sigmoid	$f(x) = \frac{1}{1+e^{-\alpha x}} \alpha > 0$	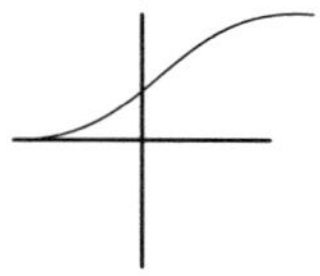
Tangent sigmoid	$f(x) = \left(\frac{2}{1+e^{-\alpha x}}\right) - 1 \ \ \alpha > 0$	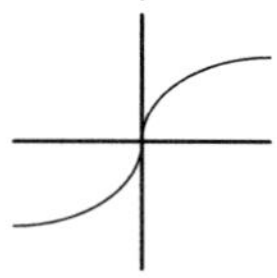
Radial basis	$f(x) = e^{-x^2/\sigma^2}$	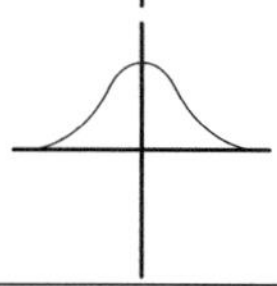

Table 5.3 shows a list of transfer functions that are usually used within the artificial neurons. Details on the role of those functions are discussed in the next sections where different networks are introduced.

Example 5.2: Dimension of Weight and Bias Matrices

It is needed to map a 3-dimensional input to a 2-dimensional output by a three-layer ANN with four neurons in the hidden layer. What would be the dimension of the weight and bias matrices?

Solution

The architecture of the network is shown in the following figure.

The weighted input in the *hidden* layer is obtained as

$$I_{4\times1} = IW'_{3\times4} \times X_{3\times1} + IB_{4\times1}$$

Fig. 5.9 Network of Example 5.2

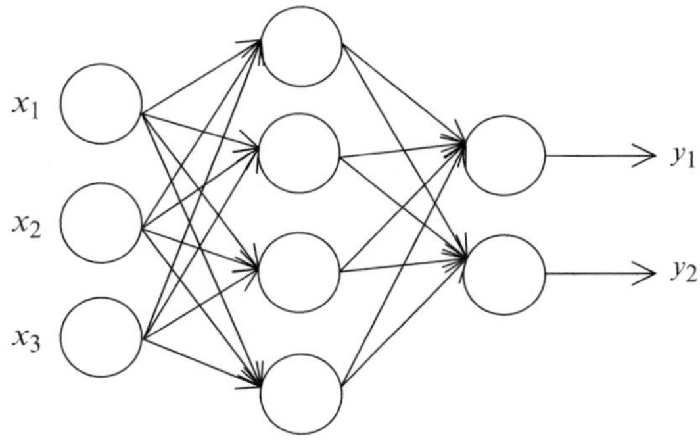

where

$$X = \begin{bmatrix} x_1 \\ x_2 \\ x_3 \end{bmatrix}; \quad IW = \begin{bmatrix} w_{11} & w_{12} & w_{13} & w_{14} \\ w_{21} & w_{22} & w_{23} & w_{24} \\ w_{31} & w_{32} & w_{33} & w_{34} \end{bmatrix}; \text{ and } \quad IB = \begin{bmatrix} b_1 \\ b_2 \\ b_3 \\ b_4 \end{bmatrix}$$

and $x_i =$ the ith input, $w_{ij} =$ the weight of the link from the ith input to the jth hidden neuron, and $b_j =$ the bias of the jth hidden neuron.

The input to the *output* layer is obtained as

$$I^o_{2\times 1} = W'^o_{4\times 2} F(I)_{4\times 1} + B^o_{2\times 1}$$

where

$$W^o = \begin{bmatrix} w^o_{11} & w^o_{12} \\ w^o_{21} & w^o_{22} \\ w^o_{31} & w^o_{32} \\ w^o_{41} & w^o_{42} \end{bmatrix}; \quad B^o = \begin{bmatrix} b^o_1 \\ b^o_2 \end{bmatrix}$$

and $F =$ the transfer function of the hidden layer.

Finally, the output matrix is obtained as

$$Y_{2\times 1} = G(I^o)_{2\times 1}$$

where $G =$ transfer function of the last layer and $Y = \begin{bmatrix} y_1 \\ y_2 \end{bmatrix}$

5.2.2 Training Algorithm

The weights and biases are parameters of a network that should be fixed before using an ANN. Weight and bias matrices of an ANN could be obtained by either supervised or unsupervised approaches. Training is an expression, which is usually termed to the supervised approach for determining weights and biases of a network. The supervised training of an ANN could be obtained by the well-known *delta* rule. The delta rule is expressed as

$$w_{ij}^{\mathrm{new}(l)} = w_{ij}^{\mathrm{old}(l)} + \eta\left(-\frac{\partial e_p}{\partial w_{ij}^{(l)}}\right) \tag{5.2}$$

where

$$E = \frac{1}{n}\sum_{p=1}^{n} e_p \tag{5.3}$$

$$e_p = \left(t_p - y_p\right)^2 \tag{5.4}$$

and n = the number of pairs of data, E = the average error of estimation, $w_{ij}^{(l)}$ = the weight of link between the ith neuron to the jth neuron in the lth layer, t_p and y_p = the target output and simulated output, respectively, and η = learning rate, the value of which is selected between 0 and 1 experimentally.

To apply the delta rule into the training process of an ANN, backpropagation (BP) algorithm is widely used. The *BP* algorithm changes the mathematical expression of the delta rule to the computational relations, which could be applied through an iterative procedure. The backpropagation algorithm, also called the generalized delta rule, provides a way to calculate the gradient of the error function efficiently using the chain rule of differentiation (Bose and Liang 1996). In this algorithm, network weights are moved along the negative of the gradient of the performance function through each iteration (which is usually called epoch) in the steepest descent direction.

For a particular weight in the lth hidden layer, the chain rule gives

$$\frac{\partial e_p}{\partial w_{ij}^{(l)}} = \frac{\partial e_p}{\partial I_{pj}^{(l)}} \cdot \frac{\partial I_{pj}^{(l)}}{\partial w_{ij}^{(l)}} \tag{5.5}$$

meanwhile

$$\frac{\partial I_{pj}^{(l)}}{\partial w_{ij}^{(l)}} = \frac{\partial}{\partial w_{ij}^{(l)}}\left(\sum_{k} w_{ij}^{(l)} y_{pi}^{(l-1)}\right) = y_{pi}^{(l-1)} \tag{5.6}$$

and $\frac{\partial e_p}{\partial I_{pj}^{(l)}}$ is defined as $-\ \delta_{pj}$. Therefore,

$$\frac{-\partial e_p}{\partial w_{ij}^{(l)}} = \delta_{pj}^{(l)} y_{pi}^{(l-1)} \tag{5.7}$$

and Eq. (5.2) is rewritten as

$$w_{ij}^{\text{new}(l)} = w_{ij}^{\text{new}(l)} + \eta \delta_{pj}^{(l)} y_{pi}^{(l-1)} \tag{5.8}$$

The chain rule is used again to define $\delta_{pj}^{(l)}$ for the last layer

$$\delta_{pj}^{(l)} = \frac{-\partial e_p}{\partial I_{pj}^{(l)}} = \frac{-\partial e_p}{\partial y_{pj}^{(l)}} \cdot \frac{\partial y_{pj}^{(l)}}{\partial I_{pj}^{(l)}} = \frac{-\partial e_p}{\partial y_{pj}^{(l)}} . f_j'(I_{pj}) \tag{5.9}$$

where

$$\frac{\partial e_p}{\partial y_{pj}^{(l)}} = -2\left(t_{pj} - y_{pj}\right) \tag{5.10}$$

Therefore, for the last layer,

$$\delta_{pj}^{(l)} = \left(t_{pj} - y_{pj}\right) f'_{Jj}(I_{pj}) \tag{5.11}$$

It should be notified that 2 could be eliminated by defining appropriate error function.

To define $\delta_{pj}^{(l)}$ for the hidden layers, the following relations are used:

$$\frac{\partial e_p}{\partial y_{pj}^{(l)}} = \sum_k \frac{\partial e_p}{\partial I_{pk}^{(l+1)}} \cdot \frac{\partial I_{pk}^{(l+1)}}{\partial y_{pj}^{(l)}} = -\sum_k \delta_{pk}^{(l+1)} w_{jk}^{(l+1)} \tag{5.12}$$

$$\delta_{pj}^{(l)} = \sum \delta_{pk}^{(l+1)} w_{jk}^{(l+1)} . f'\left(I_{pj}^{(l)}\right) \tag{5.13}$$

Using Eqs. (5.8), (5.11), and (5.13), the backpropagation algorithm is presented as follows:

1. Weights and biases are initialized as w_{ij}^{old}.
2. The output signals are generated applying input vectors (Fig. 5.10a)
   $y_{pj}^{1} = g_j[\sum w_{ij} x_{pi}]$ for the first layer.
   $y_{pj}^{(l)} = f_j\left[\sum w_{ij}^{l} y_{pi}^{(l-1)}\right]$ for the hidden layers.

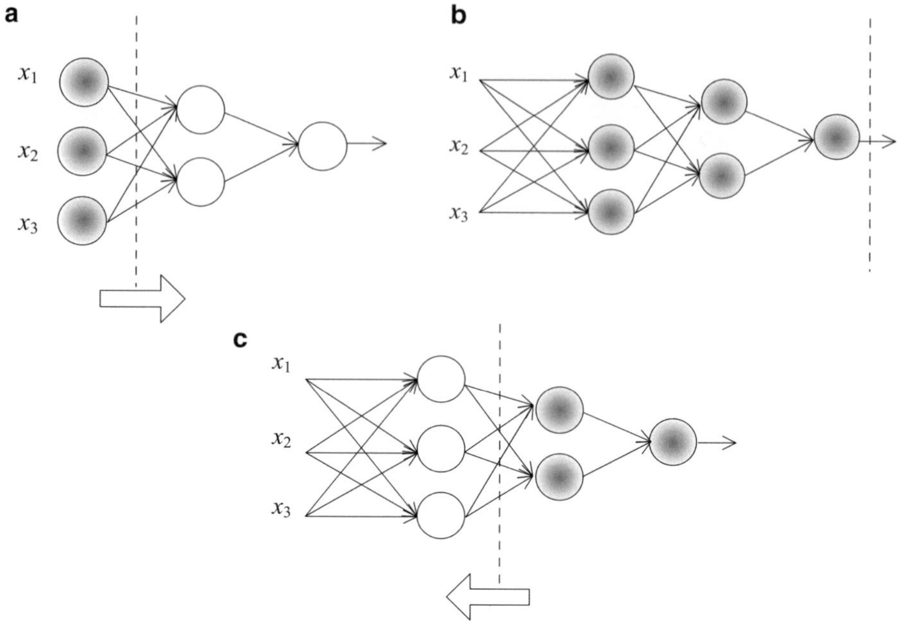

Fig. 5.10 The algorithmic steps of the BP algorithm (**a**) Feed forward simulation (**b**) Error estimation (**c**) Back propagation

3. The output signals of the last layer are compared with the targets to determine $\delta_{pj}^{(l)}$ using Eq. (5.11) (Fig. 5.10b).

$$\delta_{pj} = \left(t_{pj} - y_{pj}\right) f_j^{'}(I_{pj})$$

4. The error, δ_{pj}, is backpropagated through the output layers using Eq. (5.13) (that is why the algorithm is called backpropagation) (as shown in Fig. 5.10c).

$$\delta_{pj}^{(l)} = \left(\sum_k \delta_{pj}^{(l+1)} w_{jk}^{(l+1)}\right) f_j^{'}\left(I_{pj}^{l}\right)$$

5. Using the results of steps 3 and 4, the weights are updated in the last and hidden layers as

   $w_{ij}^{\text{new}} = w_{ij}^{\text{old}} + \eta \delta_{pj} y_{pi}^{(l-1)}$ for the last layer.

   $w_{ij}^{new(l)} = w_{ij}^{old(l)} + \eta \delta_{pj}^{(l)} y_{pi}^{(l-1)}$ for the hidden layer.

The above algorithm is continued until the desired criteria of network training are satisfied. Each iteration consisting the four latter steps of the *BP* algorithm is called *epoch*. The criteria for stopping the training procedure might be the number of epochs, minimum desired value of performance function, run time of the process, and the criteria known as *stopped training*, which is defined in the next section.

After training, the network is ready to simulate the outputs associated with the specific input vectors using the final derived weights and biases.

ANNs could be trained by either batch training or incremental training. In the batch training, weights and biases are only updated after all the inputs and targets are presented. In the incremental training the weights and biases are updated after each input is presented. In this case, the inputs and targets are presented as sequences, which are generally used in dynamic networks. In the adaptive calibration (training), there is no need to have a significant amount of data and the only essential data are the most recent ones observed before the time of simulation.

5.2.3 *Mapping by ANNs*

Function approximation is one of the most applicable uses of ANNs. The need for function approximations arises in many branches of applied mathematics and engineering in particular. In general, a function approximation problem asks us to select a function among a well-defined class that closely matches ("approximates") a target function in a task-specific way.

One can distinguish two major classes of function approximation problems: First, for known target functions approximation theory is the branch of numerical analysis that investigates how certain known functions (e.g., exponential functions) can be approximated by a specific class of functions (e.g., polynomials function) that often have desirable properties (inexpensive computation, continuity, integral and limit values, etc.). Second, in the target function, instead of an explicit formula only a set of points of the form $(x, g(x))$ is provided.

The application of an ANN as a function approximator can be divided to five steps as follows:

1. Data preprocessing
2. Selecting network architecture
3. Network training
4. Simulation
5. Postprocessing

Different algorithms and methods could be used at each step of the above process. Selection of a specific algorithm and method depends on the specific characteristics of a problem. This section discusses the frequent methods used in the mapping process by ANNs as well as the MATLAB codes developed for those purposes.

5.2.3.1 Data Preprocessing

After selecting appropriate predictors, which are actually the inputs of a mapping problem, they might be prepared before being introduced to a network. Three common processes are usually applied for this purpose as follows:

Standardizing

Standardizing helps rescaling the input values to a uniformed scale. The following command standardizes the inputs to fall in the range [−1,1]:

```
[y1,PS] = mapminmax(x1)
```

where *PS* contains the process setting parameters. The transformed matrix could be reversed to the original vector by the following command:

```
x1_again= mapminmax('reverse',y1,PS)
```

Normalizing

The following command normalizes a vector to have zero mean and unity variance, which is another way to rescale the inputs:

```
[y1,PS] = mapstd(x1)
```

The transformed matrix could be reversed to the original vector by the following command:

```
x1_again= mapstd('reverse',y1,PS);
```

Principal Component Analysis

As described in Chap. 3, the principal component analysis (PCA) detects linear dependencies between variables and replaces groups of correlated variables by new uncorrelated variables, the *principal components* (PCs). The use of PCs instead of the original input improves the mapping process by making inputs to be independent from each other and to reduce the dimension of the input vectors in case that there is a dependency between them. PCA decreases the dimension of input data by eliminating those principal components that have less contribution to the total variation in the data set. The following command extracts principal components from the input vector and eliminates those principal components that contribute less than $d\%$ to the total variation in the data set:

```
[y1,ps] = processpca(x1,d);
```

The transformed matrix is reversed to the original vector by the following command:

```
x1_again = processpca('reverse',y1,ps)
```

Before training networks, the data is usually divided into three subsets. The first subset is the training set, which is used for computing the gradient and updating the network weights and biases. The second subset is the validation set. The error on the validation set is monitored during the training process. The validation error normally decreases during the initial phase of training, as does the training set error. However, when the network begins to overfit the data, the error on the validation set typically begins to rise. The network weights and biases are saved at the minimum of the validation set error. This technique is discussed in more detail in the next pages.

The third subset is the test set. The test set error is used neither in training nor in validation. It is used to compare different models. It is also useful to plot the test set error during the training process. If the error on the test set reaches a minimum at a significantly different iteration number than the validation set error, this might indicate a poor division of the data set.

The following command divides targets into three sets using random indices:

```
[trainInd,valInd,testInd] =
dividerand(Q,trainRatio,valRatio,testRatio)
```

The following command divides targets into three sets using blocks of indices:

```
[trainInd,valInd,testInd] =
divideblock(Q,trainRatio,valRatio,testRatio)
```

The following command divides targets into three sets using specified indices:

```
[trainInd,valInd,testInd] =
divideind(Q,trainRatio,valRatio,testRatio)
```

The default ratios for all three commands are 0.7, 0.15, and 0.15 for training, validation, and test sets, respectively. Furthermore, using the following command, all targets are assigned as training set:

```
[trainInd,valInd,testInd] = dividetrain(Input)
```

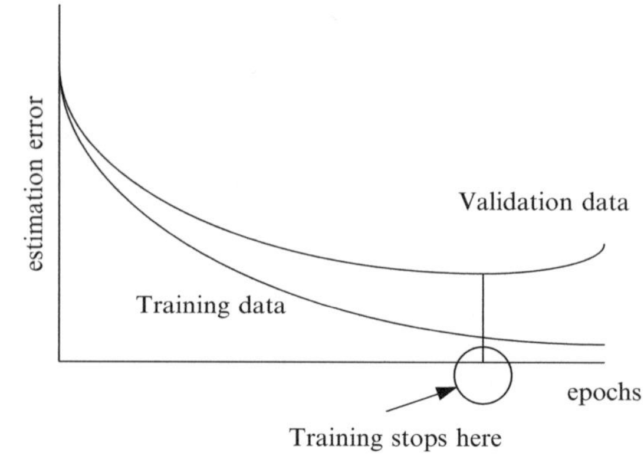

Fig. 5.11 Selection of optimum epoch based on the network performance in data training and testing

5.2.3.2 Selecting Network Architecture

A network architecture includes number of hidden layers, number of hidden neurons, specific transfer functions, the flow of data (straight or recurrent), and the way neurons are connected (say fully connected). There are well-known networks with specific architecture which are widely used in the field of water resources and environmental engineering. Multilayer perceptron (MLP), recurrent neural networks, time-delay neural networks, radial basis function networks, generalized regression neural networks, and probabilistic neural networks are examples of the well-known architectures. Those networks are presented in the next sections of this chapter.

5.2.3.3 Network Training

As far as the function mapping is concerned, training is defined as the process of calibrating the network using pairs of input/output. Artificial neural networks may suffer from the underfitting and overfitting during the training procedure (Coulibaly et al. 2000). These two factors tend to decrease the ability of the network in the generalization performance. Increasing the number of epochs in the training procedure results in decreasing the underfitting of the network, but if the number of epochs is greater than a specific number, overfitting may occur. The number of epochs is optimally determined by comparing the error in the training and testing procedure of the model. The optimal number of epochs is the number which causes the minimum validation error (Fig. 5.11).

The following command is used to train a network by specific "inputs" and "targets":

```
[net,tr]= train(net,inputs,targets);
```

5.2.3.4 Network Simulation

Simulation is actually the final aim of applying the networks. The general syntax for simulation a network is

```
a = net (inputs);
```

5.2.3.5 Postprocessing

Postprocessing includes all the tests we apply to validate the results of a specific network as well as to describe and analyze the final performance of the network. It also involves presenting ideas that might improve the performance of a network.

Three statistics are used for comparison of the results. The root-mean-square error (RMSE) is defined as

$$\text{RMSE} = \frac{\sqrt{\sum_{i}^{n} (\text{obs}_i - \text{for}_i)^2}}{n} \tag{5.14}$$

where obs_i = ith observed data, est_i = ith estimated variable, and n = number of observed values. Percent volume error (% VE) is defined as

$$\%\text{VE} = \frac{\sum_{i=1}^{n} \left| \frac{\text{obs}_i - \text{est}_i}{\text{obs}_i} \right|}{n} \tag{5.15}$$

The %VE statistic measures the absolute relative bias error of estimated values. The final statistic is the correlation (CORR) which measures the linear correlation coefficient between the observed and forecast data.

In case that the criteria is not satisfying, one can follow up some changes such as changing the number of hidden layers, changing the number of hidden neurons, changing the transfer functions, and of course changing the initial weights and biases.

5.3 Types of Artificial Neural Networks

5.3.1 *Multilayer Perceptron*

Multilayer feedforward network uses supervised training procedure that consists of providing input/output examples to the network and minimizing the error function E, which is expressed as follows:

$$E = \frac{1}{n} \sum_{p=1}^{n} \left(y_p - \hat{y}_p \right)^2 \tag{5.16}$$

where n = number of input/output data sets and $\hat{y}_p$ and y_p = observed and simulated output of the pth set, respectively. In a three-layer network with m neurons in the hidden layer, this procedure is done by utilizing the following equations:

$$I_{pj} = \sum_{i=1}^{N} w_{ji}x_{pi} + w_{jo} \tag{5.17}$$

where I_{pj} = weighted inputs into the jth hidden unit, N = total number of input nodes, w_{ji} = the weight from input unit i to the hidden unit j, x_{pi} = the value of the ith input of the pth set, w_{j0} = bias for neuron j, and g = the transition function. g is considered as a sigmoid function as follows:

$$g\left(I_{pj}\right) = \frac{1}{1 + e^{-I_{pj}}} \tag{5.18}$$

The output unit is then calculated as

$$\hat{y}_{pj} = g\left(I_{pk}\right) \tag{5.19}$$

$$I_{pj} = \sum_{j=1}^{q} w_{kj}g\left(I_{pj}\right) + w_{ko} \tag{5.20}$$

where m = number of hidden units, w_{kj} = weight connecting the hidden node j to the output k, w_{ko} = bias for neuron k, and $\hat{y}_{pj}$ = the kth estimated output for pth set. The weights in the training process are updated through each iteration in the steepest descent direction as described by *BP* algorithm. An example of architecture of an MLP has been shown in Figs. 5.9 and 5.10.

Hornick et al. (1989) theoretically proved that three-layer perceptrons with a sigmoid activation function are universal approximators, which means that they can be trained to approximate any mapping between the inputs and outputs.

The architecture of an MLP with specific "inputs" and "targets" is determined and set up by

```
net = feedforwardnet([S1 S2 ... SM]);
net = configure(net,inputs,targets);
```

where SM = number of neurons in the Mth hidden layer. The architecture of the network is viewed by

```
view(net)
```

It should be noted that the size of the input and output layers is fixed by the problem formulation. There is no systematic approach to select the number of neurons in the hidden layer. The number of neurons in the hidden layer, which produce the dependent values with more accuracy, might be selected through validation of different network architectures.

The network is trained by Levenberg–Marquardt (trainlm) as default using the following command:

```
[net,tr] = train(net,Inputs,Targets);
```

The quasi-Newton backpropagation method, *trainbfg*, and Bayesian regulation backpropagation, *trainbr*, are other options which could be used for network training.

Minimum gradient magnitude, maximum training time, minimum performance value, maximum number of validation increases, and maximum number of training epochs (iterations) could be fixed by the following command, respectively:

```
net.trainParam.min_grad=1e-10;
net.trainParam.time=60;
net.trainParam.goal=1e-5;
net.trainParam.max_fail=6;
net.trainParam.epochs=1000;
```

max_fail refers to the times validation performance has increased since the last time it is decreased.

The initial weight and bias matrices of an MLP could also be fixed by the following commands, which present an example for a 2-2-1 network:

```
net.IW{1}=[0.5 0.5; 0.5 0.5];
net.LW{2,1}=[0.5 0.5];
net.b{1}=[0.5;0.5];
net.b{2}=[0.5];
```

If the network is not accurate, it could be initialized randomly by the following command to be trained again:

```
net = init(net);
```

As described before, changing the number of hidden neurons, the training functions, and of course adding additional data for the network training is more likely to produce a network that generalizes better.

Table 5.4 Data provided for Example 5.3

P	T	P	T	P	T	P	T	P	T	P	T
1	0.50	6	0	11	−0.50	16	0.87	21	−1.00	26	0.87
2	0.87	7	−0.50	12	0.00	17	0.50	22	−0.87	27	1.00
3	1.00	8	−0.87	13	0.50	18	0.00	23	−0.50	28	0.87
4	0.87	9	−1.00	14	*0.87*	19	−0.50	24	0.00	29	0.50
5	0.50	10	−0.87	15	1.00	20	−0.87	25	0.50	30	0.00

Table 5.5 Parameters and variables used for the program of Example 5.3

Variable/ parameter	Description
AE(*S*)	Averaged simulation error associated with each "number of hidden neurons" (*S*)
bestnet	The net with the minimum error recognized by the best number of hidden neurons
E	1*30 matrix of simulation error
Emin	1*30 matrix of minimum simulation error (in percent) obtained by "*bestnet*"
nn	The maximum number of hidden neurons which is considered for the network through the program
P	1*30 matrix of input data from 1 to 30
Sbest	Number of hidden neurons associated with the minimum averaged simulation error
T	1*30 matrix of output data obtained by the selected function in association with input *P*
Y	1*30 matrix of simulated output of input *P* obtained by MLP
Ysim	Simulated outputs obtained by "*bestnet*"

Example 5.3: Mapping Mathematical Functions

Investigate the performance of MLP in mapping linear, quadratic, and sinusoidal functions.

Solution

The following program is used to determine the best number of hidden neurons for a three-layer network to map sinusoidal function.

Considering the function of $T = \sin(2\pi P/12)$, 30 values of P and T are given in Table 5.4:

The following program is used to map the sinusoidal function based on 30 data for supervising. The same program could be applied for other types of functions in case of using input/target pairs of data associated to those functions. This program fields the best number of hidden neurons from 1 to 8, which better fits the data.

```
nn=8;
minAE=10E6;
U=-99*ones(nn);
AE=U(1,:);

for S=1:nn
  net = newff(P,T,S);
  net.trainParam.epochs = 1000;
  net = train(net,P,T);
  Y = sim(net,P);
  E=abs(Y-T);
  AE(S)=mean(E);
      if AE(S)< minAE
        minAE=AE(S);
        Sbest=S;
        bestnet=net;
      end;
end;

Ysim=sim(bestnet,P);
Emin=(T-Ysim);

for SS=1:30
  Emin(SS)=Emin(SS)/T(SS)*100;
end

x=1:1:nn;
y=AE;
plot(x,y);
xlabel('number of hidden neurons');
ylabel('averaged error');
```

Table 5.5 demonstrates variables and parameters used in the program.

Figure 5.12 demonstrates the plot of "number of hidden neurons" versus averaged simulation error. According to the averaged error of simulation, the best 3-layer MLP is obtained as 1-7-1 network.

The detailed result of mapping sinusoidal function is presented here. The weights and biases of the network are shown by the commands

```
IW=bestnet.IW;
IB=bestnet.b{1};
WO=bestnet.LW{2,1};
BO=bestnet.b{2};
```

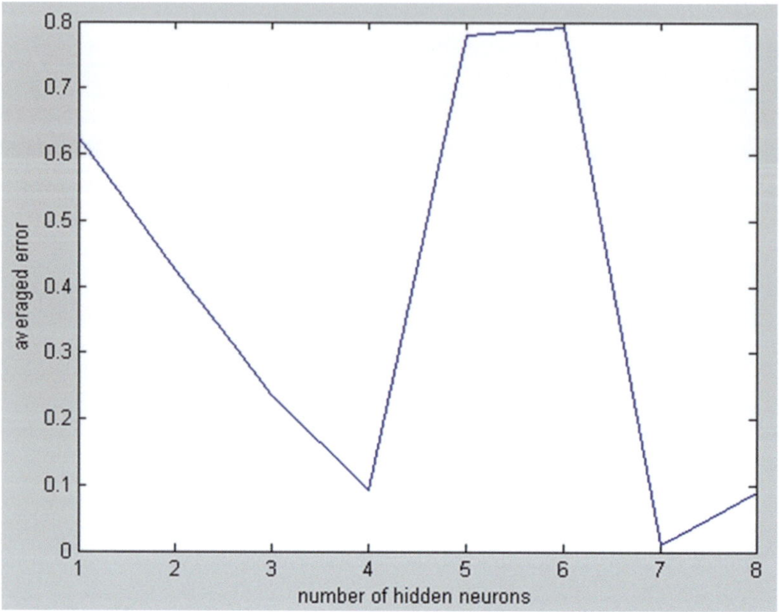

Fig. 5.12 Number of hidden neurons versus the averaged error in Example 5.3

Table 5.6 The simulation results of Example 5.3

P	*Ysim*	*P*	*Ysim*	*P*	*Ysim*	*P*	*Ysim*	*P*	*Ysim*	*P*	*Ysim*
1	0.50	6	0	11	−0.50	16	0.87	21	−1.00	26	**0.86**
2	**0.95**	7	−0.50	12	0.00	17	**0.53**	22	−0.87	27	**0.99**
3	**1.04**	8	**−0.86**	13	0.50	18	0.00	23	**−0.51**	28	0.87
4	0.87	9	−1.00	14	**0.89**	19	**−0.54**	24	0.00	29	0.50
5	0.50	10	−0.87	15	1.00	20	**−0.90**	25	**0.52**	30	0.00

which result in the following matrices:

$$IW = \begin{bmatrix} -7.68 \\ -5.90 \\ 6.52 \\ -8.19 \\ 10.13 \\ 5.59 \\ -9.94 \end{bmatrix}; \quad IB = \begin{bmatrix} 7.34 \\ 3.43 \\ -1.17 \\ -2.56 \\ 1.67 \\ 3.65 \\ -10.09 \end{bmatrix}; \quad W^o = \begin{bmatrix} 1.00 \\ -1.38 \\ -1.32 \\ -0.78 \\ 0.54 \\ -1.39 \\ -0.97 \end{bmatrix}; \quad B^o = [-0.61]$$

Finally the simulated results are shown in Table 5.6.

The performance of the network could also be shown by

```
plotperf(tr)
```

Table 5.7 Data presented for Example 5.4

P1	P2	*T*	P1	P2	*T*
24	110	487	42	28	320
26	75	351	48	18	300
32	105	483	105	45	389
35	65	329	58	22	340
45	25	350	120	18	311
50	20	300	20	100	439
100	40	359	30	80	379
54	20	320	32	105	483
123	16	309	60	25	320
30	70	339	110	18	291

Please note that due to the random selection of the initial weights and biases in this example, the results might change at each run of the program. In case of considering similar initial weights, the results remain constant at each run.

Example 5.4: Using Cross-Validation Approach in Network Training

For the input data given in Table 5.7, find the best number of neurons between 2 and 3 in a three-layer MLP that fits the following data.

Solution

The following program uses a cross-validation approach to assess the performance of a 2-2-1 MLP in mapping data. In the cross-validation approach, once each pair of input/output is omitted from the *n* observation of the data and the other *n-1* pairs of data are used to estimate the omitted one. This is repeated *n* times changing the omitted pair of data and the averaged simulation error for all *n* pairs of data are considered as the indicator of the real performance of the network.

```
%<<<<<<<<<<<<<< Pre-processing >>>>>>>>>>>>>>
[pn,ps1] = mapstd(P);
[ptrans,ps2] = processpca(pn,0.02);
PC=ptrans;
%<<<<<<<<<<<<<< Pre-processing >>>>>>>>>>>>>>
for n=1:1:20

%<<Generation of data sets for cross validation>>
```

(continued)

```
    Ptest=[PC(1,n);PC(2,n)];
    Ttest=T(n);
     for m=1:1:19
       if m<n
       B(1,m)= PC(1,m);
       B(2,m)=PC(2,m);
       TB(m)=T(m);
       else
         B(1,m)=PC(1,m+1);
         B(2,m)=PC(2,m+1);
         TB(m)=T(m+1);
       end
     end
%<<<<<Network architecture and configuration >>>>>
net                    =                    newff(B,TB,[2],
{'tansig','purelin'},'trainlm',...
'learngdm','mse',
{'fixunknowns','removeconstantrows','mapminmax'}...
,{'removeconstantrows','mapminmax'},'divideblock');
%<<<<< Initial weights and biases >>>>>>
   net.IW{1}=[0.5 0.5; 0.5 0.5];
   net.LW{2,1}=[0.5 0.5];
   net.b{1}=[0.5;0.5];
   net.b{2}=[0.5];
%<<<<<<<<<<<<<< Training >>>>>>>>>>>>>>
   net.trainParam.epochs = 1000;
   net = train(net,B,TB);
%<<<<<<<<<<<<<< Simulation >>>>>>>>>>>>>>
   Y = sim(net,Ptest);
   S(n)=Y;
%<<<<<<<<<<<<<< Post-processing >>>>>>>>>>>>>>
   E(n)=abs(Y-Ttest);
end

%<<<<<<<<<<<<<<     Averaged     network     error
>>>>>>>>>>>>>>

AE=mean(E);
```

Table 5.8 Parameters and variables used in Example 5.4

Variable/parameter	Description
B	Input matrix for calibration
E	Matrix of simulation error
P	Matrix of input data
PC	Principal components of the standardized data
Ptest	Input matrix for validation
T	Target matrix
TB	Target matrix for calibration
Ttest	Target matrix for validation
Y	Simulated outputs

To assess the 2-3-1 network, only the following two changes are made in the network architecture and initial weights and biases:

```
newff(B,TB,[3],{'tansig','purelin'},'trainlm',...

'learngdm','mse',{'fixunknowns','removecon-
stantrows',
  'mapminmax'}...
```

and

```
net.IW{1}=[0.5 0.5; 0.5 0.5;0.5 0.5];
   net.LW{2,1}=[0.5 0.5 0.5];
   net.b{1}=[0.5;0.5;0.5];
   net.b{2}=[0.5];
```

Table 5.8 demonstrates variables and parameters used in the program.

Obviously, 2-3-1 network outperforms 2-2-1. Figure 5.13 demonstrates the observed versus simulated data by two networks.

5.3.2 Dynamic Neural Networks

Static neural networks such as multilayer perceptron network (MLP) only process input patterns that are spatial in nature, i.e., input patterns that can be arranged along one or more spatial axes such as a vector or an array. In many tasks, the input pattern comprises one or more temporal signals, as in speech recognition, time

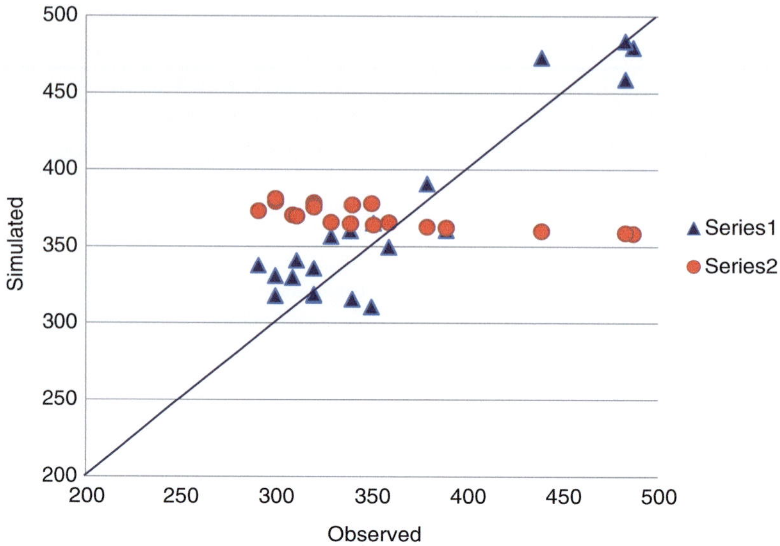

Fig. 5.13 Results obtained for Example 5.4

Fig. 5.14 Schematic of a tapped delay line

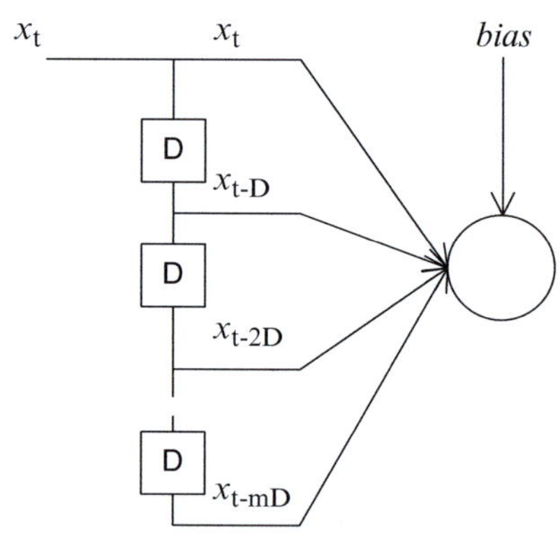

series prediction, and signal filtering (Bose and Liang 1996). Tapped delay lines (TDLs) and recurrent (feedback) connections are two components, which could be assigned for static networks in order to design temporal neural networks. The networks, which apply either a TDL or recurrent connections, are referred to as dynamic neural networks.

A tapped delay line (TDL) consists of several time-delay operators, which are arranged in an incremental order. A typical TDL is shown in Fig. 5.14. As shown in this figure, a time-delay operator, D, is a memory box, which receives an input signal at each time step and saves it along one time step. After passing one time interval, the operator results the signal as an output. As shown in Fig. 5.9, the input signal enters from the left and passes through $N - 1$ delays. The output of a *TDL* is an N-dimensional vector, made up of the input signal at the current time and the previous input signal. Tapped delay lines explicitly represent a temporal process. Replacing the internal connection weights in an MLP network by the tapped delay lines results in a network called time-delay neural network (TDNN) (Waibel et al. 1989; Atiya and Parlos 1992; Wan 1993).

Another component for involving initial and past states of a system is recurrent (feedback) connection. This connection recurs from either the output layer or the hidden layer back to a context unit and after one time step returns to the input layer. A context unit consists of several time-delay operators, which represent dimension of time implicitly by its effects on the processing. It is in contrast to a TDL component, which explicitly considers temporal processing. Attaching recurrent connections (context unit) to an MLP network results the recurrent neural network (RNN). There are different models of RNN, depending on the architecture of the recurrent connections: the Jordan RNN (Jordan 1986), which has feedback connections from the output layer to input layer; Elman RNN (Elman 1990) which has feedback connections form hidden layer to input layer.

Because dynamic networks have memory, they can be trained to learn sequential or time-varying patterns. In dynamic networks, the output depends not only on the current input to the network but also on the previous inputs, outputs, or states of the network. Although dynamic networks can be trained using the same gradient-based algorithms that are used for static networks, the performance of the algorithms on dynamic networks can be quite different, and the gradient must be computed in a more complex way.

Some of the recent developments to take into account the temporal characteristics into the neural networks are creating a spatial representation of temporal pattern by considering a sliding window of input sequences (Hsu et al. 1995; Jain et al. 1999; Coulibaly et al. 2000), putting time delays (tapped delay lines) into the neurons or their connections (Waibel et al. 1989; Wan 1993; Sajikumar and Thandaveswara 1999; Karamouz et al. 2004), employing recurrent connections (Jordan 1986; Elman 1990), and using combination of the above-mentioned methods (Coulibaly et al. 2001).

5.3.2.1 Input Delay Neural Network

The most straightforward dynamic network is called the input delay neural network (IDNN) also known as focused time-delay neural network (FTDNN). This network consists of an MLP with a tapped delay line at the input layer as shown in Fig. 5.14. Figure 5.15 illustrates a two-layer IDNN.

The FTDNN network is created using *timedelaynet* command. This command is similar to the *feedforwardnet* command, with the additional input of the tapped delay line vector.

```
ftdnn_net = timedelaynet([0:D],[S1 S2 ... SM]);
```

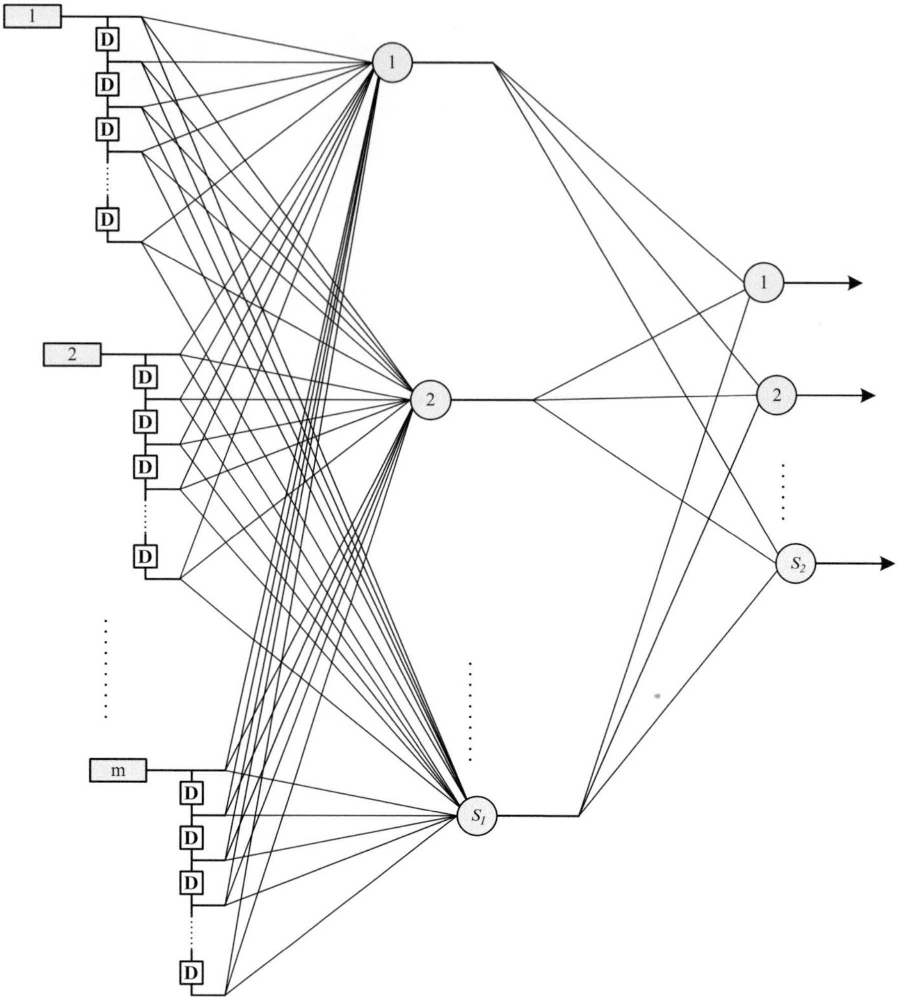

Fig. 5.15 Schematic of an IDNN

where D is the maximum time delay for the input series.

ANNs can be trained either statically or adaptively. In order to model time series, the adaptive (recursive) training method is an appropriate alternative. The recursive (also called online) training allows the recalibration and training of the model as soon as new observations become available. This dynamic adaptability enables the model parameters to be adjusted to the properties of an ongoing event by capturing the characteristics of the current situation (Toth et al. 2000). Subsequently, it could fully utilize the capacity of the neural networks for time series modeling.

Example 5.5: Time Series Modeling

The annual inflow to a reservoir is tabulated in Table 5.9. Use an IDNN to forecast one-step-ahead inflow to the reservoir.

Solution
The following program is used for this example.
First the numerical array is changed to cell array.

```
x = num2cell(P);
```

```
ftdnn_net = timedelaynet([1:2],10);
ftdnn_net.trainParam.epochs = 3000;
ftdnn_net.divideFcn = '';
P = x(3:end);
T = x(3:end);
Pi=x(1:2);
ftdnn_net = train(ftdnn_net,P,T,Pi);
Y = ftdnn_net(P,Pi);
e = gsubtract(Y,T);
rmse = sqrt(mse(e));
```

Table 5.9 The inflow data for Example 5.5

Year	Inflow	Year	Inflow	Year	Inflow	Year	Inflow
1957	1,440.8	1972	1,461.1	1987	1,524.4	2002	1,451.9
1958	1,442.6	1973	1,581.4	1988	1,466.2	2003	1,456.3
1959	1,365.2	1974	1,459.8	1989	1,460.9	2004	1,450.0
1960	1,406.9	1975	1,526.0	1990	1,424.6	2005	1,475.1
1961	1,365.1	1976	1,434.6	1991	1,559.1	2006	1,465.7
1962	1,413.2	1977	1,443.8	1992	1,464.4		
1963	1,435.4	1978	1,424.7	1993	1,482.2		
1964	1,397.6	1979	1,523.7	1994	1,540.8		
1965	1,390.0	1980	1,472.3	1995	1,501.6		
1966	1,386.7	1981	1,451.5	1996	1,465.5		
1967	1,430.9	1982	1,450.8	1997	1,502.3		
1968	1,692.7	1983	1,441.0	1998	1,432.7		
1969	1,449.7	1984	1,457.2	1999	1,413.0		
1970	1,444.5	1985	1,457.4	2000	1,405.9		
1971	1,530.1	1986	1,463.5	2001	1,442.9		

Table 5.10 Variables and parameters used in the program of Example 5.5

Variable/parameter	Description
E	Series of prediction error
P	Array of the inflow to the reservoir
RSME	Root-mean-square error of prediction
X	Cell array of the inflow to the reservoir
Y	One-step-ahead predicted time series

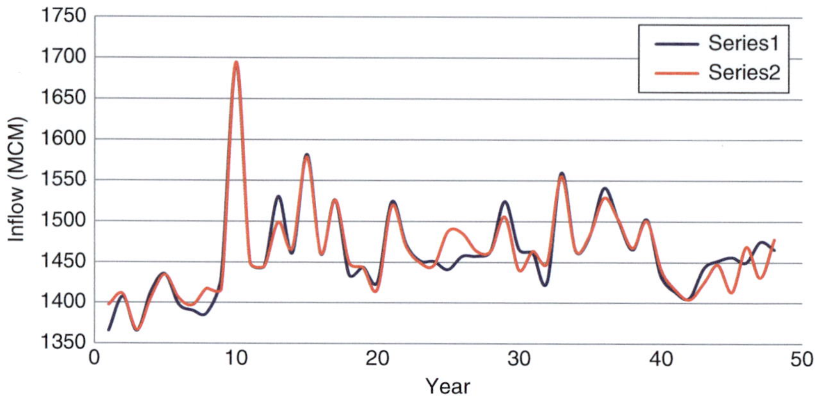

Fig. 5.16 The results of Example 5.5

Because the network has a tapped delay line with a maximum delay of *2*, the prediction is begun by the (*2* + *1*)th value of the time series. It is also needed to load the tapped delay line with the *2* initial values of the time series (contained in the variable *Pi*). Notice that the input to the network is the same as the target. Because the network has a minimum delay of one time step, this means that a one-step-ahead prediction is performed.

Table 5.10 demonstrates the variables and parameters used in the program.

The predicted time series (series 2) versus actual series (series 1) is shown in Fig. 5.16.

5.3.2.2 Time-Delay Neural Network

The IDNN had the tapped delay line memory only at the input to the first layer of the MLP network. The tapped delay lines could be considered throughout the hidden layers of the network, which is called distributed TDNN. Figure 5.17 shows a general two-layer distributed TDNN.

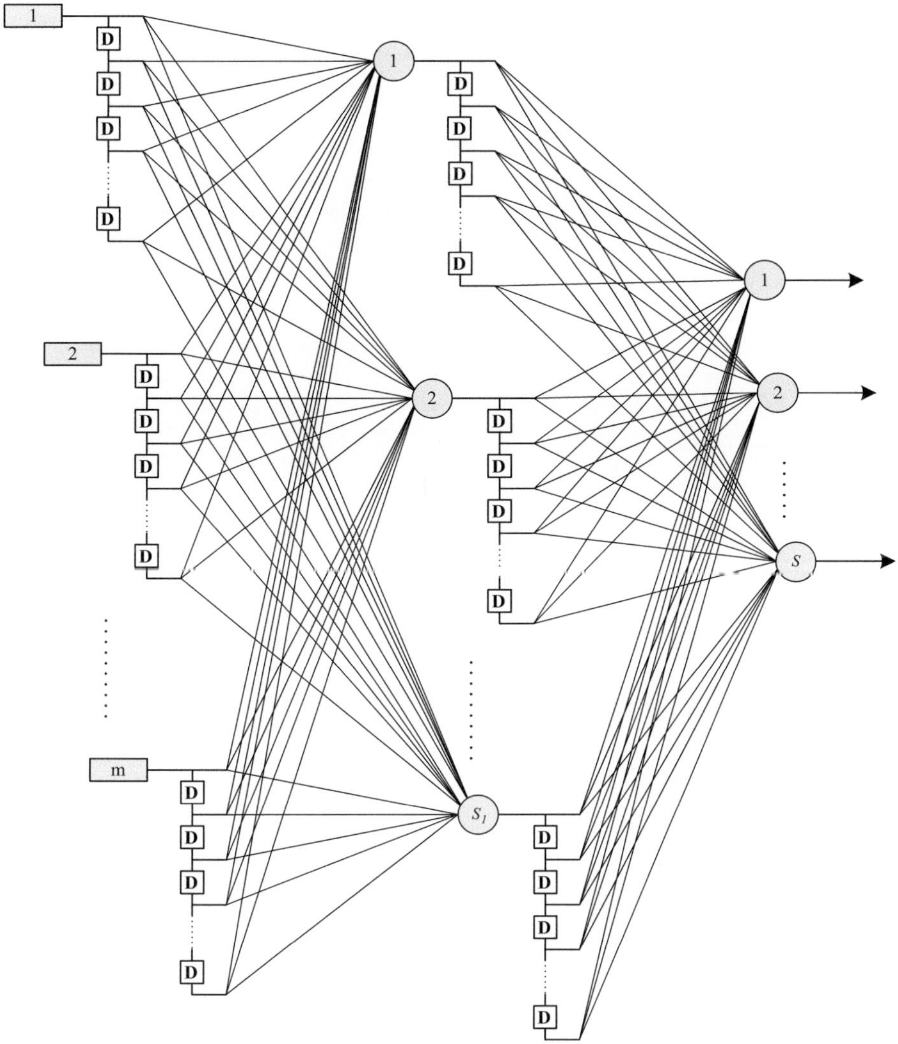

Fig. 5.17 Schematic of a time-delay neural network

The outputs of each layer are obtained as

$$y_j^1(t) = F\left(\sum_{d_1=0}^{D_1}\sum_{i=1}^{m} w_{i,j,d}^1 p_{i,d+1}(t) + b_j^1\right) \quad 1 \le j \le S_1 \tag{5.21}$$

$$y_k^2(t) = G\left(\sum_{d_2=0}^{D_2}\sum_{j=1}^{S_1} w_{j,k,d_2}^2 y_{j,d_2}^1(t) + b_k^2\right) \quad 1 \le k \le S_2 \tag{5.22}$$

where $y_k^2(t)$ and $y_k^2(t)$ = output from the first and the second layers, respectively; F and G = transfer function of the first and the second layers, respectively; w and b = weights and biases, respectively; S_1 and S_2 = number of neurons in the first and second layers, respectively; D_1 and D_2 = time delay in the first and the second layers, respectively; and p = input data.

The distributed TDNN network is created by the *distdelaynet* function. The only difference between the *distdelaynet* function and the *timedelaynet* function is that the first input argument is a cell array that contains the tapped delays to be used in each layer.

The time-delay neural network is created using the following command:

```
d1 = 0:D1;
d2 = 0:D2;
dtdnn_net = distdelaynet({d1,d2},[S1 S2 ... SM]);
```

where DM is the maximum delay considered for the Mth layer.

Example 5.6: Time Series Modeling by TDNN

Solve Example 5.5 by a TDNN network.

Solution

The approach is the same as Example 5.6. The program is changed to the following form:

```
x=num2cell(P);
```

```
d1=1:2;
d2=1:2;
P = x(3:end);
T = x(3:end);
Pi=x(1:2);
dtdnn_net = distdelaynet({d1,d2},9);
dtdnn_net = train(dtdnn_net,P,T,Pi);
Y = sim(dtdnn_net,P,Pi);
e = gsubtract(Y,T);
rmse = sqrt(mse(e));
```

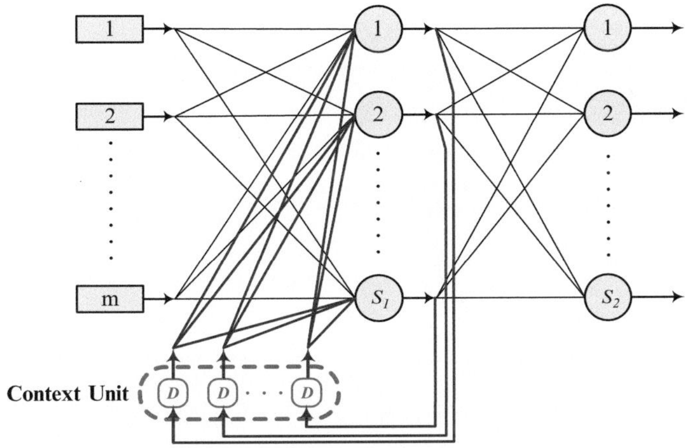

Fig. 5.18 Schematic of a layer-recurrent network

5.3.2.3 Layer-Recurrent Network

A layer-recurrent network contains a feedback loop, with a single delay, around each layer of the network except for the last layer (Fig. 5.18). The first version of this network was introduced by Elman (1990). The *newlrn* command develops an Elman network as

```
lrn_net = newlrn(P,T,S);
```

where S = number of neurons in the hidden layer.

5.3.2.4 The Nonlinear Autoregressive Network with Exogenous Inputs

The nonlinear autoregressive network with exogenous inputs (NARX) is a recurrent dynamic network, with feedback connections enclosing several layers of the network. The NARX model is based on the linear ARX model, which has been described in time series modeling (Chap. 3).

In a NARX network, the output is fed back to the input of the feedforward neural network as shown in Fig. 5.19. This output could be considered as either the estimated output or the actual output. Using the actual output has two advantages. The first is that the input to the feedforward network is more accurate. The second is that the resulting network has a purely feedforward architecture, and static backpropagation can be used for training. The syntax of developing a NARX model is

```
d1 = [1:D1];
d2 = [1:D2];
narx_net = narxnet(d1,d2,S);
```

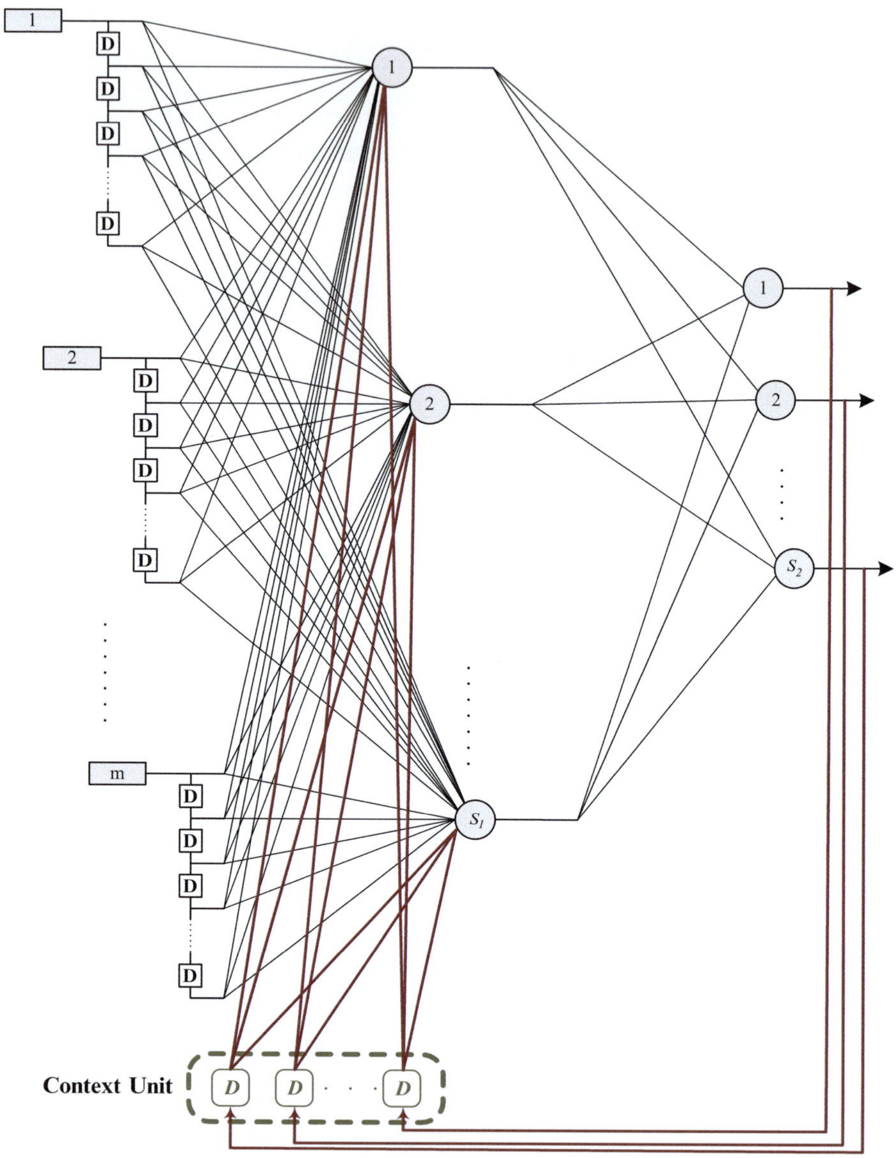

Fig. 5.19 Schematic of a TDRNN

where D_1 and D_2 = time delay of the input and output connections, respectively, and S = number of hidden neurons.

To convert the network to use the feedback of the simulated output instead the actual output, the following command is used:

```
narx_net_closed= closeloop(narx_net);
```

Example 5.7: Rainfall–Runoff Modeling (Flood Prediction)

Table 5.11 shows a 4-h excess rainfall as well as a 1.5-h excess rainfall and the consequent flood hydrographs. Set up a network to predict flood hydrograph of 2-h storm in the region (Table 5.12, Figs. 5.20 and 5.21).

Solution

It is assumed that the flood hydrograph (yp) of excess rainfall, R, could be simulated by the following function:

$$yp(t) = f[R(t), R(t-1), yp(t-1)]$$

After training NARX, to map the above function, a closed loop version of NARX is used for prediction of flood values $\hat{yp}(t)$ according to the following function:

$$\hat{yp}(t) = f[R(t), R(t-1), \hat{yp}(t-1)]$$

Table 5.11 The calibration data in Example 5.7

Time (0.5 h)	Rainfall (mm)	Hydrograph (cm)	Time (0.5 h)	Rainfall (mm)	Hydrograph (cm)
1	12.7	5.7	17	0	0.0
2	19.0	23.8	18	0	0.0
3	25.4	67.5	19	0	0.0
4	38.1	132.9	20	50.8	22.9
5	63.5	214.5	21	76.2	95.4
6	50.8	307.0	22	25.4	235.7
7	31.7	403.8	23	0	371.3
8	25.4	446.3	24	0	361.7
9	0	396.8	25	0	220.5
10	0	297.2	26	0	101.3
11	0	191.7	27	0	60.7
12	0	105.6	28	0	43.8
13	0	54.0	29	0	22.4
14	0	30.3	30	0	4.9
15	0	13.9	31	0	0.0
16	0	4.9	32	0	0.0

Table 5.12 The validation data in Example 5.7

Time (0.5 h)	Rainfall (mm)
1	26.9
2	49.0
3	62.2
4	46.0

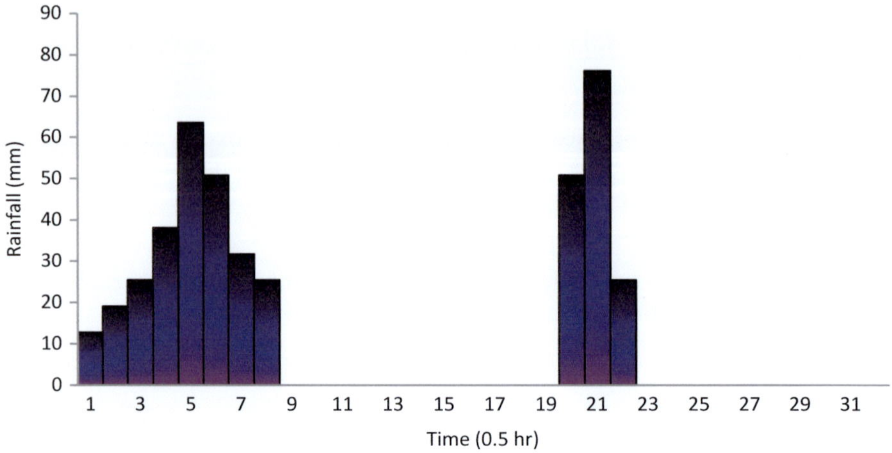

Fig. 5.20 Rainfall data in Example 5.7

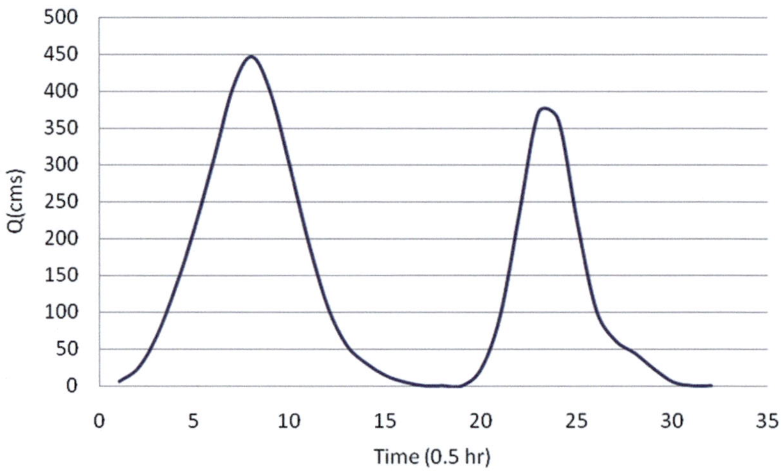

Fig. 5.21 Hydrograph data in Example 5.7

The following program is developed for this purpose:

```
R=num2cell(R);
F=num2cell(F);
RR=num2cell(RR);

%<<<<< Trainng >>>>>
d1 = [0:1];
d2 = [1:1];
narx_net = narxnet(d1,d2,9);
narx_net.trainparam.epochs=3000;
```

(continued)

```
narx_net.trainParam.min_grad = 1e-10;
[p,Pi,Ai,t] = preparets(narx_net,R,{},F);
narx_net = train(narx_net,p,t,Pi);
yp = sim(narx_net,p,Pi);
e = cell2mat(yp)-cell2mat(t);
plot(e)


%<<<<<< Prediction >>>>>
narx_net_closed = closeloop(narx_net);
[pp,PPi] = preparets(narx_net_closed,RR);
ypp = sim(narx_net_closed,pp,PPi);
```

with the following variables and parameters (Table 5.13).

Figure 5.22 shows the predicted (red line) versus the observed hydrograph (blue line).

Table 5.13 Variables and parameters used in the program of Example 5.7

Variable/ parameter	Description
E	Simulation error for the observed hydrographs
F	1*32 matrix of the observed flood hydrograph values as shown in the table
R	1*32 matrix of observed rainfall hyetograph values as shown in the table
RR	1*12 matrix of the rainfall hyetograph values with unknown consequent flood. The rainfall values after the fourth time interval are considered as zero
Yp	Simulated observed hydrographs
Ypp	Simulated unknown hydrograph

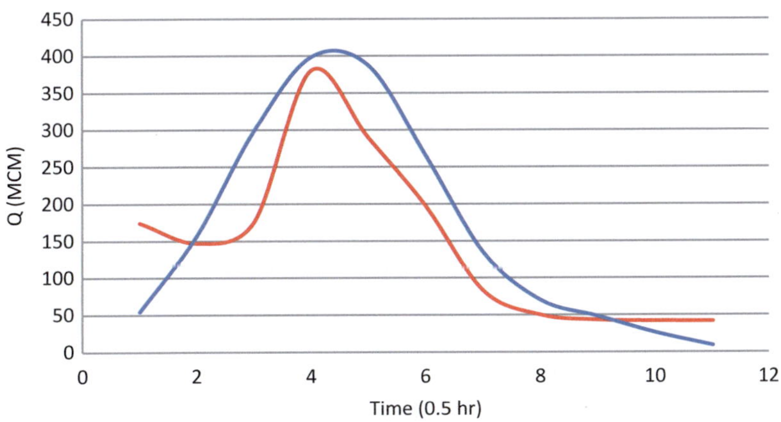

Fig. 5.22 The results of hydrograph simulation in Example 5.7

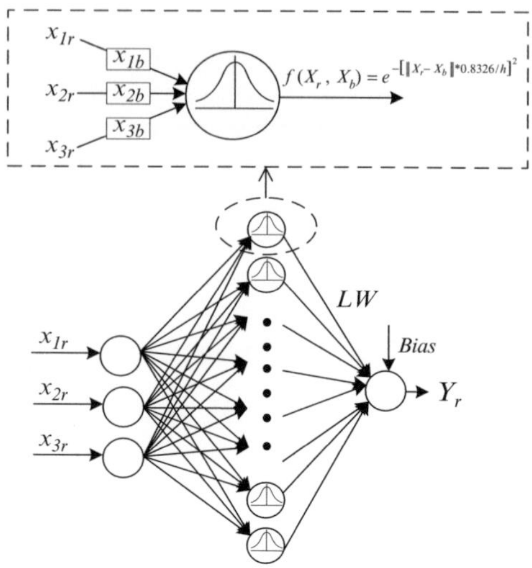

Fig. 5.23 Schematic of an RBF network

5.3.3 *Statistical Neural Networks*

5.3.3.1 Radial Basis Function

Radial basis function (RBF) is the basis for radial basis networks, which are in turn the basis for series of networks known as statistical neural networks. Statistical neural networks are referred to as the networks which in contrast to the conventional neural networks use regression-based methods and are not inspired by the biological neural system (Picton 2000). An RBF network uses the similarity between observations of predictors and similar sets of historical observations (successors) to obtain the best estimate for a dependent variable. An RBF is a three-layer network, with only one hidden layer (Fig. 5.23). The number of neurons in the hidden layer is equal to the number of historical observation of predictors (successors). In fact, each neuron in the hidden layer represents a pair of historical observation of predictors/dependants. The output of each neuron is actually the contribution of the historical observation in estimating the real-time event.

A typical RBF neuron is shown in Fig. 5.13. As it is shown in this figure, RBF uses a Gaussian performance function. The input to this function is the Euclidian distance between each input to the neuron and the specified vector of the same size of the input. The Gaussian function uses the following relation:

$$
\begin{aligned}
f(X_r, b) &= e^{-I^2} \\
I &= \|X_r - X_b\| * 0.8326/h
\end{aligned}
\tag{5.23}
$$

where X_r = network input with unknown output, X_b = observed inputs in time or location b, and h = spread. The output of the function approaches 0 to 1, when $\|X_r - X_b\|$ approaches a large value to 0, respectively. The value of the output between those limits depends on h, which is also known as *spread*.

The general form of calculating a dependent variable (Y_r) by predictor X_r is then

$$Y_r = \mathrm{LW} * f(X_r, X_b) + \mathrm{Bias} \tag{5.24}$$

where LW and Bias = weight matrix of connections from the hidden layer to the output layer and bias matrix of the output layer, respectively. When an RBF network is developed, LW and bias matrices are calculated by solving the system of equations of $T_b = \mathrm{LW} * f(X_r, b) + \mathrm{Bias}$ where T_b is the target associated with the bth observation.

For a set of input/outputs, X and Y, an RBF is set up as

```
net = newrbe(P,T,h);
```

In contrast to the MLP network, there is no need to train RBF. Actually RBF is ready to simulate while it is set up. The network is used for simulation by the following command:

```
Ysim = sim(net,Psim);
```

Example 5.8: Spatial Interpolation

Table 5.14 and Fig. 5.24 show the coordinates of measuring rainfall stations at a field. Use the recorded annual rainfall data to estimate rainfall at non-measuring locations shown in Fig. 5.24.

The annual rainfall in this example is considered as a function of the coordinates. Therefore, the input/output data would be the coordinates and rainfall depth, respectively:

```
P= [3 4 5 6 7 2 1 0 9 10 10 0 9 1; 2 4 1 7 8 9 1 2 4 5 0 9 10 7];
T=[300 350 250 500 500 50 20 150 400 400 10 50 70 20];
```

The question is "what is the best *spread* for interpolation of rainfall in the field?" We use a try-and-error approach to examine different values of *spread* starting from 0.1 ending to 6 with 0.1 incremental steps. It should be noted that since the radial basis network plays the role of an exact estimator function, the application of different *spread* values within the calibration set results in a similar averaged

Table 5.14 The information of the stations in Example 5.8

Stations	Longitude	Latitude	Annual rainfall
A	3	2	300
B	4	4	350
C	5	1	250
D	6	7	500
E	7	8	500
F	2	9	50
G	1	1	20
H	0	2	150
I	9	4	400
J	10	5	400
K	10	0	10
L	0	9	50
M	9	10	70
N	1	7	20
X1	2	2	?
X2	2	7	?
X3	8	3	?

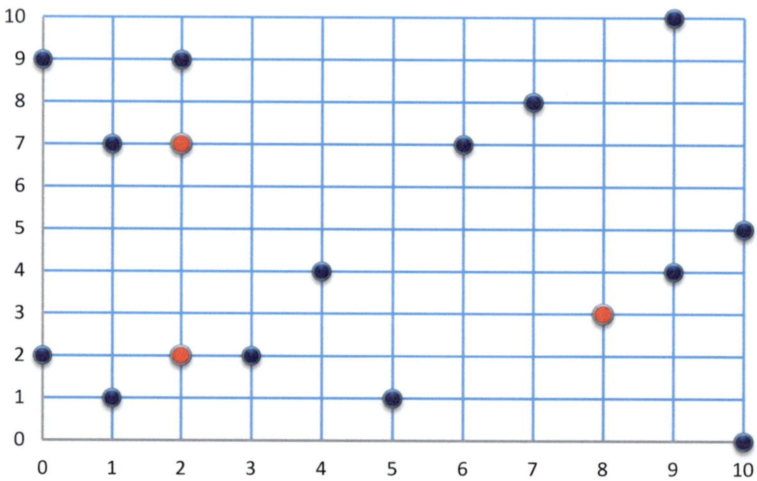

Fig. 5.24 Location of the measuring stations (*blue circles*) and non-measuring points (*red circles*)

error of approximately zero. Therefore, to examine the performance of the network in a practical manner, the approach of cross-validation is used. In this approach, each pair of input/output is omitted from the n observation of data set once and the other $n - 1$ pairs of data are used to estimate the omitted one. This iteration is repeated n times and the averaged simulation error for all n pairs of data is considered as the indicator of the real performance of the network.

The following algorithmic steps use the described approach to find out which *spread* (h) minimizes the average error of spatial rainfall estimation (Fig. 5.25).

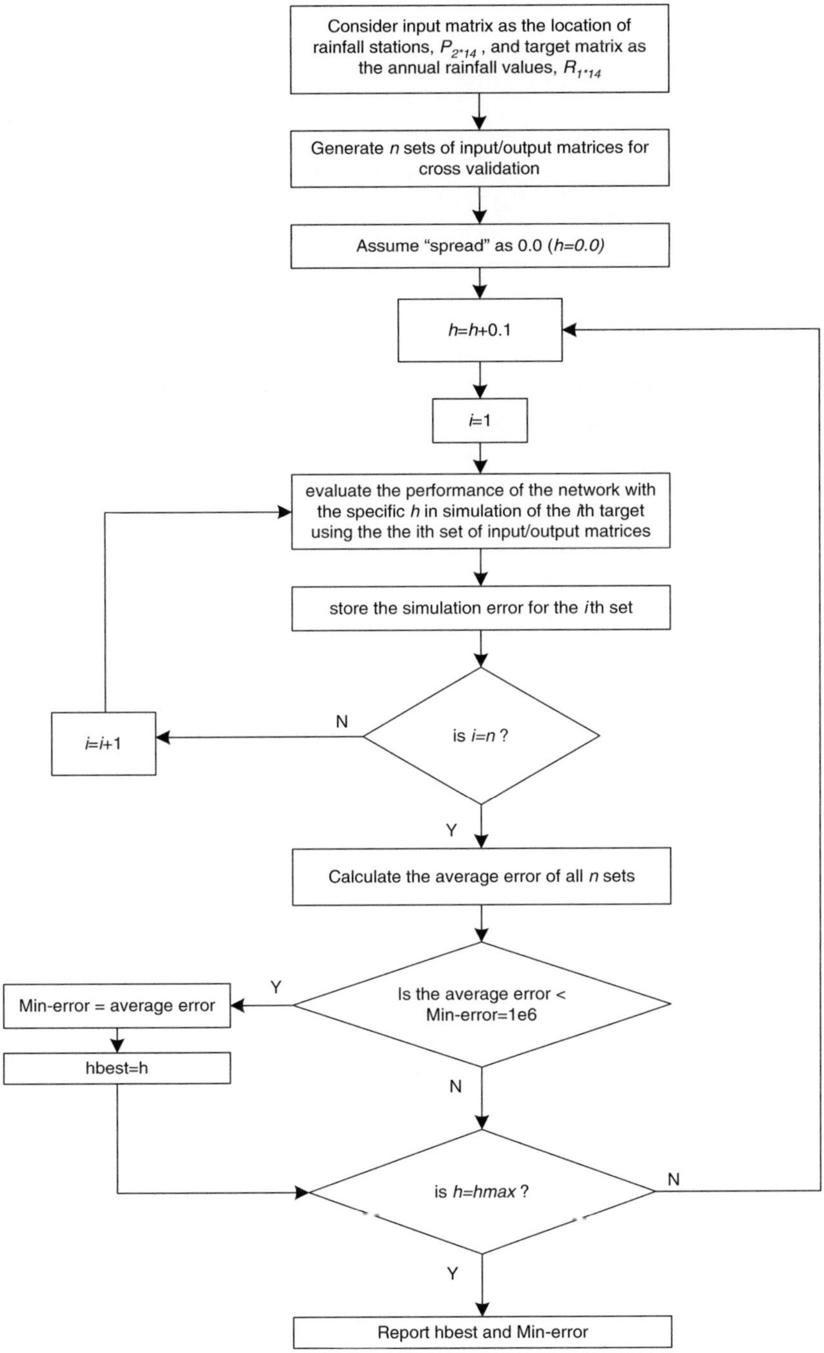

Fig. 5.25 Presented algorithm for solving Example 5.8

The program corresponding to the above algorithm is presented as follows:

```
minAE=1e+6;

for h=0.1:0.1:3
  for n=1:1:14
    TestP=[P(1,n);P(2,n)];
    TestT=[T(n)];
     for m=1:1:13
        if m<n
        B(1,m)= P(1,m);
        B(2,m)=P(2,m);
        TB(m)=T(m);
        else
          B(1,m)=P(1,m+1);
          B(2,m)=P(2,m+1);
          TB(m)=T(m+1);
        end
     end
    net = newrbe(B,TB,h);
    Y= sim(net,TestP);
    E(n)=abs(Y-TestT);
  end
 AE = mean(E);
 j=int32(h*10)
 Error(j)=AE;
  if AE < minAE
  minAE=AE;
  hbest=h;
 end
end;
h = 0.1:.1:3;
plot(h,A);
xlabel('spread');
ylabel('error');
```

The description of the notation is demonstrated by Table 5.15.

Using the above program, the best spread is obtained as 1.6 which results in an average error of almost 121 mm (Fig. 5.26). Using this value of spread, the estimated rainfall in locations *X1*, *X2*, and *X3* are 193, 356, and 87 mm, respectively.

Table 5.15 Variables and parameters used in the program of Example 5.8

Variable/parameter	Description
AE	Averaged error of interpolation associated with the specific spread (*h*)
B	2*13 matrix of coordinates for calibration
H	Spread of the network
Hbest	The best spread for interpolating rainfall in the field
TB	1*13 matrix of recorded annual rainfall for calibration
P	2*14 matrix of coordinates
T	1*14 matrix of recorded annual rainfall
TestP	2*1 matrix of coordinates for validation
TestT	1*1 matrix of recorded annual rainfall for validation

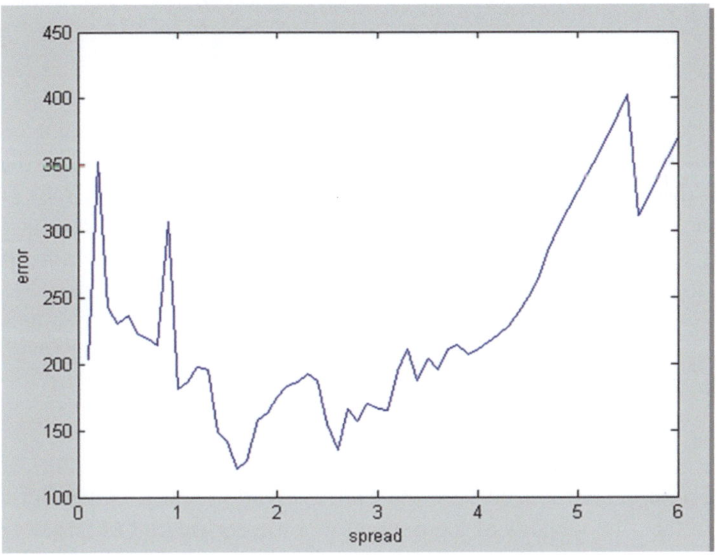

Fig. 5.26 The error of simulation versus values of spread in Example 5.8

5.3.3.2 Generalized Regression Neural Network

A generalized regression neural network (GRNN) is a variant of the radial basis function (RBF) network. GRNN is a universal approximator for smooth functions, which is able to solve any smooth function approximation problem given enough data. The *GRNN* considers every experienced pair of data as a possible class which is probably to happen regarding to the observation of specific conditions.

Figure 5.27 shows the architecture of a GRNN. *GRNN* is a three-layer network where in contrast to the *MLP*, the number of neurons in its hidden layer could be

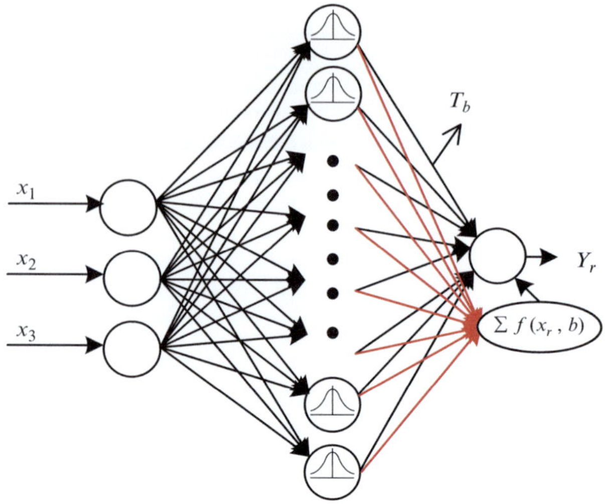

Fig. 5.27 Schematic of generalized regression neural network

understood easily as it is equal to the number of observed data. Also like the most of networks, the number of neurons in the input and output layers of *GRNN* is equal to the dimension of input and output vectors, respectively. A *GRNN* uses the same equations as the RBF network in its first layer; however, it uses the following equation to calculate an output:

$$Y_r = \frac{1}{\sum_{b=1}^{n} f(X_r, b)} \sum_{b=1}^{n} [f(X_r, b) \times T_b] \tag{5.25}$$

where T_b = target associated with the bth observation and n = number of observations. In fact, the weights of the output layer are considered as target values.

Example 5.9: Spatial Interpolation by GRNN

Solve the previous example by the use of generalized regression neural network.

Solution

The algorithm and program which used in this example are as the same as what was presented in the previous example with just a change in the network configuration as

```
net =newgrnn(B,TB,h);
```

Using the above program, the best *spread* is obtained as 0.1 which results in an average error of 65 mm (Fig. 5.28).

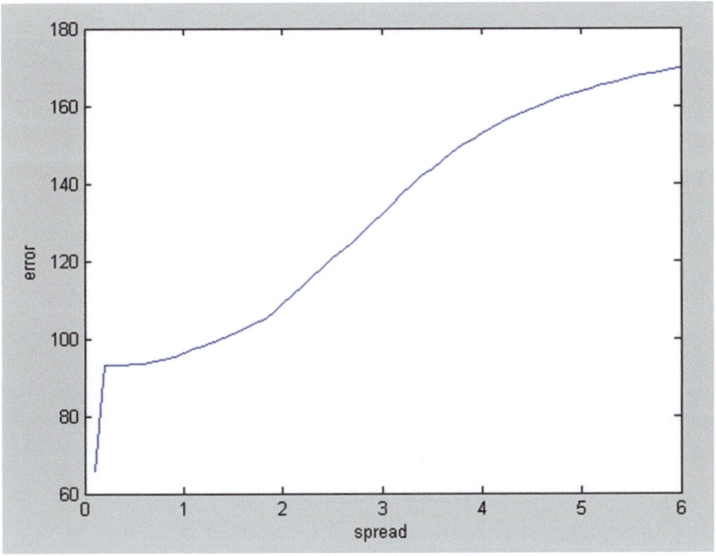

Fig. 5.28 The error of simulation versus values of spread in Example 5.9

5.3.3.3 Probabilistic Neural Network

Probabilistic neural networks can be used for classification problems. The architecture of this network is similar to that of RBF network. When an input is presented, the first layer computes distances from the input vector to the calibration input vectors and produces a vector of probabilities as $f(X_r, X_b)$. In the last layer, a *compete* transfer function on the output picks the maximum of these probabilities. The architecture for this network is shown in Fig. 5.29.

The command for using a probabilistic neural network is presented here:

```
net = newpnn(P,T,spread)
```

with the same variables presented before.

Example 5.10: Modeling Water Quality Index

The class of water quality in a region based on the nitrate (NO_3) and chlorine (Cl) are tabulated as follows. These classes have been obtained by the CCME Water

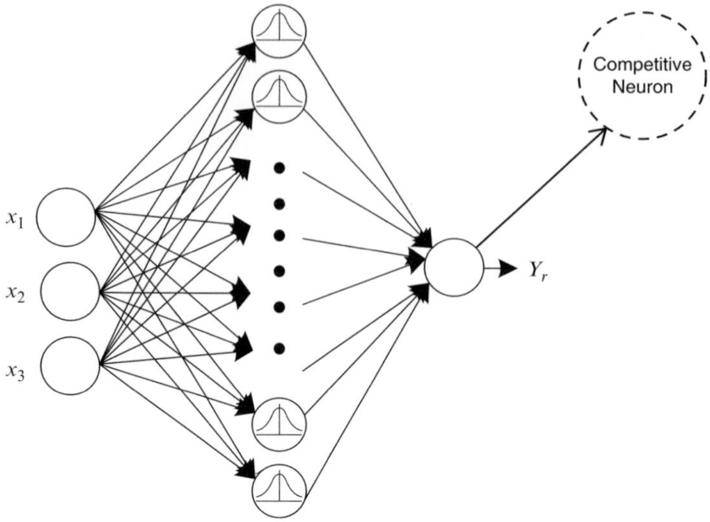

Fig. 5.29 Schematic of probabilistic neural network

Quality Index (CCME 2001). We need to set up a network to classify water quality in this region by the use of NO_3 and Cl as inputs (Table 5.16).

Solution
Thirty pairs are considered as calibration data set and the remaining 14 data are used for network validation.

Please note that the indices (targets) should be converted to vectors by the following command:

```
T= ind2vec(Tc)
```

The vector has a dimension of $m*n$, where $m =$ number of indices and n is number of observations. The simulated vector is then converted to indices by

```
Tc= vec2ind(T)
```

Table 5.16 The water quality data used for Example 5.10

Cl(mg/l)	No3(mg/l)	CCME-WQI	Number of class	Class
41	44.8	100	5	Excellent
14	10	100	5	Excellent
40	17.7	100	5	Excellent
88	104	46.53	2	Marginal
88	134.5	41.13	1	Poor
74	80.1	52.98	2	Marginal
68	11.4	100	5	Excellent
126	28.8	100	5	Excellent
38	16.2	100	5	Excellent
146	21.6	100	5	Excellent
245	25.5	100	5	Excellent
350	62	59.63	2	Marginal
500	19.5	61.92	2	Marginal
63	53.4	62.88	2	Marginal
48	65.2	58.38	2	Marginal
74	42	100	5	Excellent
52	57.6	61.4	2	Marginal
74	58.6	61.05	2	Marginal
59	55	61.41	2	Marginal
60	71	56.13	2	Marginal
102	141	40.28	1	Poor
186	1.5	100	5	Excellent
70	7.1	100	5	Excellent
96	81	52.69	2	Marginal
188	38	100	5	Excellent
84	47.6	64.41	3	Fair
70	47.2	64.48	3	Fair
114	67.5	57.5	2	Marginal
76	79.5	53.13	2	Marginal
72	29.6	100	5	Excellent
105	75.8	54.5	2	Marginal
82	143.6	39.94	1	Poor
174	134.5	41.13	1	Poor
82	58.4	61.05	2	Marginal
98	27.4	100	5	Excellent
70.2	47	64.53	3	Fair
59.4	45.8	64.61	3	Fair
72	59	60.88	2	Marginal
64.8	72.5	55.63	2	Marginal
37.8	35.8	100	5	Excellent
75.7	86.5	51.02	2	Marginal
64	68	57.33	2	Marginal
44.6	25.1	100	5	Excellent
62	70.8	56.3	2	Marginal

To employ PNN, first it is needed to find out what is the best *spread* for a specific classification. The following program uses cross-validation to select the best spread:

```
minAE=10000;

for h=2:2:30
  for n=1:1:30
    TestP=[P(1,n);P(2,n)];
    TestT=[T(n)];
     for m=1:1:29
       if m<n
        B(1,m)= P(1,m);
        B(2,m)=P(2,m);
        TB(m)=T(m);
       else
         B(1,m)=P(1,m+1);
         B(2,m)=P(2,m+1);
         TB(m)=T(m+1);
       end
     end
    TT = ind2vec(TB);
    net = newpnn(B,TT,h);
    Y = sim(net,TestP);
    Yc = vec2ind(Y);
    E(n)=abs(Yc-TestT);
  end
  AE = mean(E);
  j=int32(h)
  Error(j)=AE;
  if AE < minAE
  minAE=AE;
  hbest=h;
 end
end;
Ytest = sim(net,Ptest);
Yctest = vec2ind(Ytest);
```

The description of the variables and parameters is demonstrated by Table 5.17.

The best spread for classification is obtained as 14. The validation results are presented in Table 5.18.

Table 5.17 Variables and parameters used in the program of Example 5.10

Variable/parameter	Description
AE	Averaged error of classification
B	2*29 matrix of pollutants for calibration
h	Spread of the network
hbest	The best spread for classification
TB	1*29 matrix of water quality class for calibration
P	2*30 matrix of pollutants
Ptest	2*14 matrix of pollutants for test
T	1*30 matrix of water quality class
TestP	2*1 matrix of pollutants for validation
TestT	1*1 matrix of water quality class for validation
Yctest	Simulated index of the water quality class
Ytest	Simulated vector of the water quality class

Table 5.18 The results of Example 5.10

Actual class	Simulated class	Actual class	Simulated class
2	2	2	2
1	1	2	2
1	1	5	5
2	2	2	2
5	5	2	2
3	2	5	5
3	2	2	2

5.4 Summary

The chapter presented and discussed different types of models, namely, multilayer perceptron which is categorized as static networks, dynamic neural networks, and statistical neural networks. Each category contains different networks which are briefly presented as follows.

Multilayer feedforward network uses supervised training procedure that consists of providing input/output examples to the network and minimizing the error function. This is the most widely used network in different fields of water resources and environmental engineering. Static neural networks such as multilayer perceptron network (MLP) only process input patterns that are spatial in nature, i.e., input patterns that can be arranged along one or more spatial axes such as a vector or an array. In many tasks, the input pattern comprises one or more temporal signals, as in time series prediction. Tapped delay lines (TDLs) and recurrent (feedback) connections are two components, which could be assigned for static networks in order to design temporal neural networks. The networks, which apply either a TDL or recurrent connections, are referred to as dynamic neural networks.

The IDNN had the tapped delay line memory only at the input to the first layer of the MLP network. The tapped delay lines could be considered throughout the hidden layers of the network, which is called distributed TDNN. A layer-recurrent network contains a feedback loop, with a single delay, around each layer of the network except for the last layer.

The nonlinear autoregressive network with exogenous inputs (NARX) is a recurrent dynamic network, with feedback connections enclosing several layers of the network. The NARX model is based on the linear ARX model, which has been described in time series modeling.

Radial basis function (RBF) is the basis for radial basis networks, which are in turn the basis for series of networks known as statistical neural networks. Statistical neural networks are referred to as the networks which in contrast to the conventional neural networks use regression-based methods and are not inspired by the biological neural system. An RBF network uses the similarity between observations of predictors and similar sets of historical observations (successors) to obtain the best estimate for a dependent variable. An RBF is a three-layer network, with only one hidden layer.

A generalized regression neural network (GRNN) is a variant of the radial basis function (RBF) network. GRNN is a universal approximator for smooth functions, which is able to solve any smooth function approximation problem given enough data. The *GRNN* considers every experienced pair of data as a possible class which is probably to happen regarding to the observation of specific conditions.

GRNN is a three-layer network where in contrast to the *MLP*, the number of neurons in its hidden layer could be understood easily as it is equal to the number of observed data. Also like the most of networks, the number of neurons in the input and output layers of *GRNN* is equal to the dimension of input and output vectors, respectively. A *GRNN* uses the same equations as the RBF network in its first layer; however, it uses the following equation to calculate an output.

Probabilistic neural networks can be used for classification problems. The architecture of this network is similar to that of RBF network. When an input is presented, the first layer computes distances from the input vector to the calibration input vectors and produces a vector of probabilities. In the last layer, a *compete* transfer function on the output picks the maximum of these probabilities and produces 1 for the class associated with the calibration vector input with maximum probability and 0 for the other classes.

Workshop

Zayandeh-rud River is the main surface resource for irrigation and domestic demands in the central part of Iran, especially the Isfahan metropolitan area. As

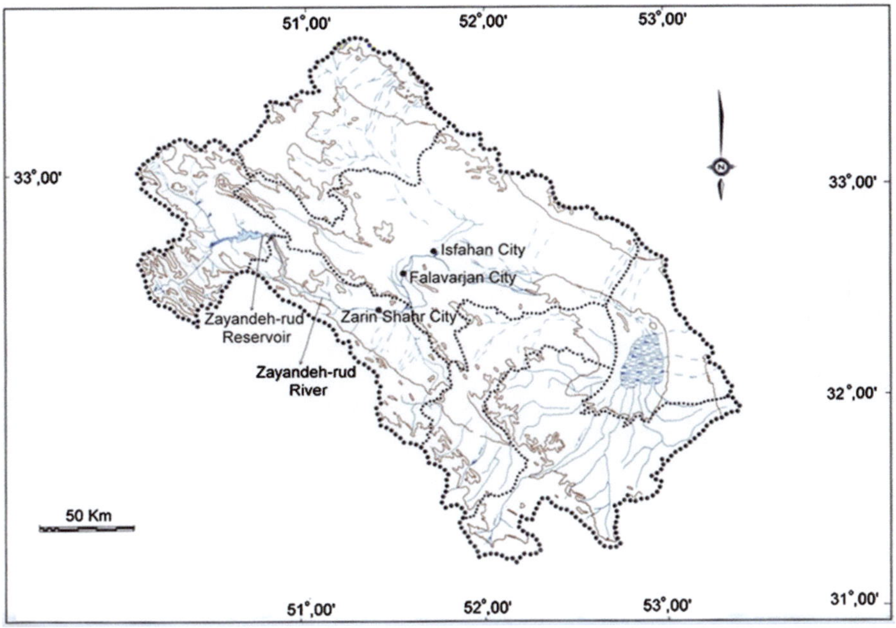

Fig. 5.30 Map of *Zayandh-rud* basin

water demands increase in Isfahan, water withdrawals from the river increase and it is critical that climate variability is incorporated into water resources-related decision-making. Zayandeh-rud reservoir (Fig. 5.30), with 1,470 million cubic meters volume, controls streamflow upstream of Isfahan city. This workshop aims to set up an appropriate model based on ANNs to forecast winter inflow (inflow from January to March) to Zayandeh-rud reservoir. Data of 30-year period from 1971 to 2002 are used in this workshop. The predictors of the winter inflow include "October to December streamflow," "averaged Southern Oscillation Index (SOI) from October to December," and "rainfall from October to December" (Table 5.19).

Solution

This workshop uses a combination of ANN-based models instead of relying on an individual network. The main idea of the approach of this case is to apply different networks, each trained to perform better in specific conditions. Actually, a network is trained to simulate dry conditions (below average streamflow) and another is trained for wet seasons (below average streamflow). To switch between these two networks in real-time forecasting, a classifying network (say PNN) could be applied to decide if a condition is likely to be either a dry or a wet season. This approach is summarized in the following diagram (Fig. 5.31).

Table 5.19 Predictors and streamflow data of the workshop

	Predictor 1	Predictor 2	Predictor 3	Dependant variable
Year	October to December streamflow	Averaged SOI from October to December	October to December rainfall	January to March streamflow
1971	181.37	13.60	472.50	–
1972	112.91	−9.77	281.70	187.43
1973	136.15	18.27	184.50	346.06
1974	146.79	6.47	394.00	201.75
1975	241.02	18.00	356.00	223.89
1976	368.61	−0.07	231.00	376.61
1977	210.26	−12.30	637.00	293.98
1978	150.44	−2.47	163.00	449.25
1979	210.85	−1.93	245.10	302.30
1980	189.13	−3.50	128.70	228.88
1981	208.71	1.70	237.00	359.68
1982	156.69	−24.23	445.00	207.64
1983	130.35	4.47	227.00	207.03
1984	121.48	0.30	337.00	193.94
1985	193.93	−2.27	416.00	188.46
1986	230.64	−4.33	613.00	168.35
1987	200.10	−6.07	414.00	412.90
1988	191.19	18.57	553.00	489.07
1989	127.92	3.67	630.00	278.49
1990	159.57	−3.70	70.00	346.09
1991	184.65	−12.27	554.00	198.91
1992	290.65	−7.90	451.00	220.49
1993	397.95	−6.83	361.00	478.63
1994	123.48	−12.87	1,041.00	330.08
1995	130.89	1.07	99.50	323.51
1996	104.09	3.67	116.00	197.57
1997	111.51	−15.93	315.00	133.38
1998	118.04	11.50	98.00	296.57
1999	81.66	7.27	385.00	180.56
2000	119.30	14.00	380.00	139.65
2001	137.06	2.23	350.00	151.39
2002	–	–	–	324.80

The following program is provided to train an MLP for dry or wet seasons. The difference between this MLP and a conventional MLP is actually in the "performance function." Where a conventional MLP uses the well-known performance function of $\sum_{p=1}^{n} (y_p - \hat{y}_p)^2$, the following modified MLP uses $\sum_{p=1}^{n} (y_p - \hat{y}_p)^2 \times 1.2^{(31-p)}$ which is applied through the commands used in

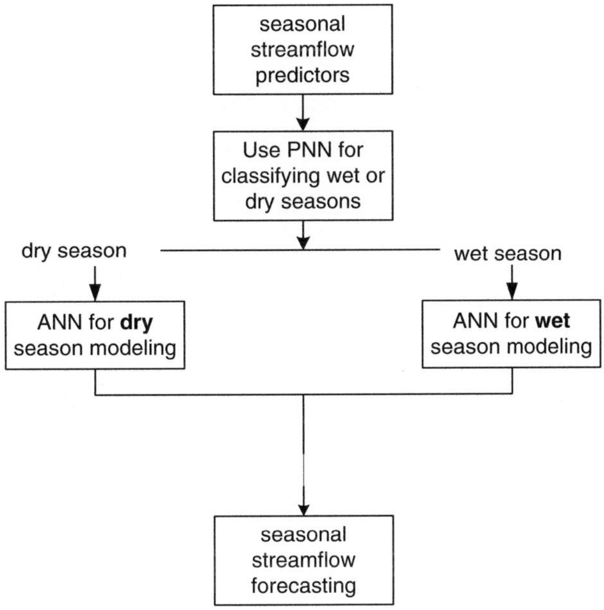

Fig. 5.31 The algorithm of the proposed method

section "Network Training" of the program. To stress on dry seasons, the pairs of input/target are sorted in an ascending order of targets and vice versa:

```
nn=20;
minAE=10E6;
U=-99*ones(nn);
AE=U(1,:);
%
%<<<<< Pre-Processing:>>>>>
%<<<<< Standardize data and transform them to
principal
%components >>>>>
%
[pn,ps1] = mapstd(P);
[ptrans,ps2] = processpca(pn,0.2);
Input=ptrans;
%
%<<<<< divide Inputs to train and validation sets
>>>>>
%
[trai-Ind,valind,testInd]=divideblock
(Input,0.8,0.2,0.0);
```

(continued)

```
for S=1:nn
%
%<<<<< Network Architecture:
%Three-layer MLP with S hidden neurons >>>>>
%
net = feedforwardnet([S]);
net = configure(net,Input,T);
%
%<<<<< Network Training >>>>>
%
net = init(net);
ind = 1:31;
ew = 1.2.^(31-ind);
net.trainParam.epochs = 1000;
net.trainParam.goal = 1e-5;
[net,tr]=train(net,Input,T,{},{},ew);
%
%<<<<< Simulation >>>>>
%
a=net(Input);
%
%<<<<< Post-Processing >>>>>
%
E=abs(a-T);
AE(S)=mean(E);
      if AE(S)<minAE
          minAE=AE(S);
          Sbest=S;
          bestnet=net;
      end

end;

Tsim=bestnet(Input);
plot(Input,T,'o',Input,Tsim,'x')
```

Comparison of the simulation obtained by the dry season model by what is obtained by a conventional MLP model is shown in the following figure (Fig. 5.32).

Also, comparison of the simulation obtained by the wet season model by what is obtained by a conventional MLP model is shown in Fig. 5.33.

A PNN program, similar to what is presented in Example 5.10, could be employed to classify dry ad wet conditions.

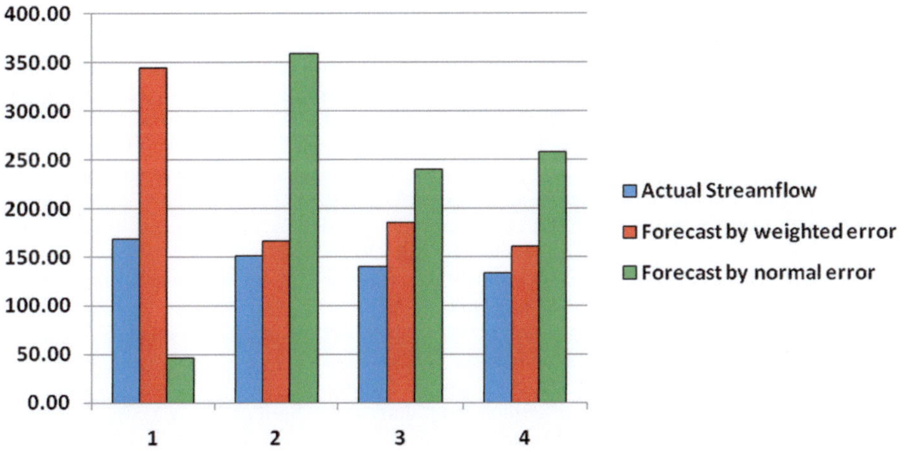

Fig. 5.32 Comparison of the simulation obtained by the dry season model by what is obtained by a conventional MLP model

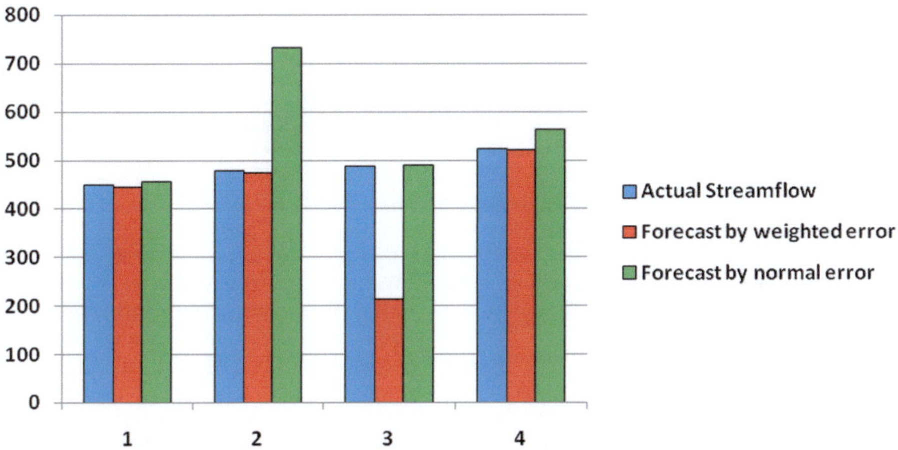

Fig. 5.33 Comparison of the simulation obtained by the wet season model by what is obtained by a conventional MLP model

References

Anderson J, Rosenfeld E (1988) Neurocomputing: foundations of research. MIT Press, Cambridge, Collection of fundamental original papers

Anmala J, Zhang B, Govindaraju RS (2000) Comparison of ANNs and empirical approaches for predicting watershed runoff. J Water Resour Plann Manage 126(3):156–166

Atiya A, Parlos A (1992) Nonlinear system identification using spatiotemporal neural networks. In: Proceedings of the international joint conference on neural networks, Baltimore, MD, vol 2. The IEEE, Inc., Piscataway, pp 504–509

Bose NK, Liang P (1996) Neural network fundamentals with graphs, algorithms, and application. McGraw-Hill, New York

CCME (Canadian Council of Ministers of the Environment) (2001) Canadian water quality guidelines for the protection of aquatic life: CCME Water Quality Index 1.0, technical report. In: Canadian environmental quality guidelines, 1999. Canadian Council of Ministers of the Environment, Winnipeg

Chandramouli V, Raman H (2001) Multireservoir modeling with dynamic programming and neural networks. J Water Resour Plann Manage 127(2):89–98

Coulibaly P, Anctil F, Bobee B (2000) Daily reservoir inflow forecasting using artificial neural networks with stopped training approach. J Hydrol 230:244–257

Coulibaly P, Anctil F, Bobee B (2001) Multivariate reservoir inflow forecasting using temporal neural networks. J Hydrol Eng ASCE 6(5):367–376

Elgaali E, Garcia LA (2007) Using neural networks to model the impacts of climate change on water supplies. J Water Resour Plan Manage 133(3):230–243

Elman JL (1990) Finding structure in time. Cogn Sci 14:179–211

Ertel W, Black N (2011) Elements of artificial neural networks. Springer, New York

Garbrecht JD (2006) Comparison of three alternative ANN designs for monthly rainfall-runoff simulation. J Hydrol Eng 11(5):502–505

Hornick K, Stinchcombe M, White H (1989) Multilayer feedforward networks are universal approximators. Neural Netw 2(5):359–366

Hsu KL, Gupta HV, Sorooshian S (1995) Artificial neural network modeling of rainfall-runoff process. Water Resour Res 31(10):2517–2530

Jain SK, Das D, Srivastava DK (1999) Application of ANN for reservoir inflow prediction and operation. J Water Resour Plan Manage ASCE 125(5):263–271

Jordan MI (1986) Attractor dynamics and parallelism in a connectionist sequential machine. In: 8th annual conference of the Cognitive Science Society. MIT Press, Amherst, pp 531–546

Karamouz M, Razavi S, Araghinejad S (2004) Application of artificial neural networks in flood estimation. In: 4th international conference on decision making in urban and civil engineering, Porto, Portugal

Kisi O, Yıldırım G (2007) Adaptive neurofuzzy computing technique for evapotranspiration estimation. J Irrig Drain Eng 133(4):368–379

Milot J, Rodriguez MJ, Serodes JB (2002) Contribution of neural networks for modeling trihalomethanes occurrence in drinking water. J Water Resour Plann Manage ASCE 128 (5):370–376

Neelakantan TR, Pundarikanthan NV (2000) Neural network-based simulation-optimization model for reservoir operation. J Water Resour Plann Manage 126(2):57–64

Picton P (2000) Neural networks, 2nd edn. Palgrave, New York

Sajikumar N, Thandaveswara BS (1999) Non-linear rainfall-runoff model using artificial neural network. J Hydrol 216:32–35

Schmid BH, Koskiaho J (2006) Artificial neural network modeling of dissolved oxygen in a wetland pond: the case of Hovi, Finland. J Hydrol Eng 11(2):188–192

Suen JP, Eheart W (2003) Evaluation of neural networks for modeling nitrate concentrations in rivers. J Water Resour Plann Manage ASCE 129(6):505–510

Thirumalaiah K, Deo MC (2000) Hydrological forecasting using neural networks. J Hydrol Eng 5 (2):180–189

Toth E, Brath A, Montanari A (2000) Comparison of short-term rainfall prediction models for real-time flood forecasting. J Hydrol 239:132–147

Trajkovic S, Todorovic B, Stankovic M (2003) Forecasting of reference evapotranspiration by artificial neural networks. J Irrig Drain Eng 129(6):454–457

Waibel A, Hanazawa T, Hintin G, Shikano K, Lang KJ (1989) Phoneme recognition using time delay neural networks. IEEE Trans ASSP 37(3):328–339

Wan EA (1993) Time series prediction using a connectionist network with internal delay lines. In: Weigend AS, Gershenfeld NA (eds) Time series prediction: forecasting the future and understanding the past. Addison-Wesley, Reading, pp 195–217

Chapter 6
Support Vector Machines

Abstract Classifying data is a common task in data-driven modeling. Using support vector machines, we can separate classes of data by a hyperplane. A support vector machine (SVM) is a concept for a set of related supervised learning methods that analyze data and recognize patterns, used for classification and regression analysis. The formulation of SVM uses the structural risk minimization principle, which has been shown to be superior to the traditional empirical risk minimization principle used by conventional neural networks. This chapter presents principles of classification and regression analysis by support vector machines, briefly. Also related MATLAB programs are presented.

Keywords Support vector machines • Classification • Kernel function • Regression

6.1 Introduction

Classifying data is a common task in data-driven modeling. Using support vector machines, we can separate classes of data by a hyperplane. A support vector machine (SVM) is a concept for a set of related supervised learning methods that analyze data and recognize patterns, used for classification and regression analysis. Support vector machine was developed by Vapnik in 1995 (Vapnik 1995). The formulation of SVM uses the structural risk minimization principle, which has been shown to be superior to the traditional empirical risk minimization principle used by conventional neural networks (Burges 1998). In what follows, the application of SVM will be introduced, briefly (Fig. 6.1).

S. Araghinejad, *Data-Driven Modeling: Using MATLAB® in Water Resources and Environmental Engineering*, Water Science and Technology Library 67,
DOI 10.1007/978-94-007-7506-0_6, © Springer Science+Business Media Dordrecht 2014

6.2 Support Vector Machines for Classification

As it was presented in Chap. 5, models such as artificial neural network divide the decision space by a line, a plane, or a hyperplane as it is presented in Fig. 6.1. The difference between SVM and other classifiers is to divide the decision space in a way that the risk of classification is minimized. It means that two classes have maximum distance from both lines. As it is demonstrated in Fig. 6.1 *a*, classifier line might be inappropriate to divide a two-dimensional space into two classes (line *a* in Fig. 6.1). Line *b* shown in Fig. 6.2 divides the space in a proper manner; however, it has a high risk of misclassification in case of adding new data points. Among the others, line *c* divides the space with the associated risk which is minimized as possible. Line *b* represents a classifier line obtained by neural networks where line *c* may represent a classifier line resulted by SVM.

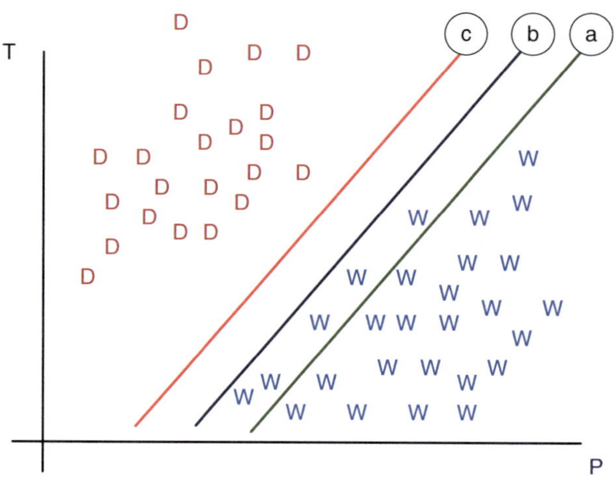

Fig. 6.1 Different examples of classification for a wrong classifier (*a*), an unstable classifier (*b*), and a stable classifier (*c*)

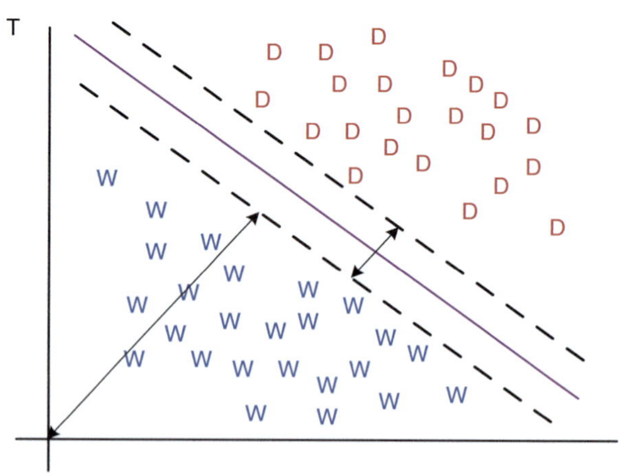

Fig. 6.2 Schematic view of classification in the SVM approach

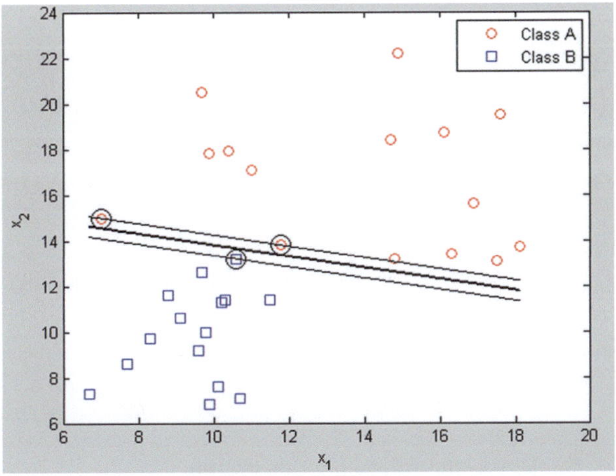

Fig. 6.3 Plot of the classification results in Example 6.1

There are many hyperplanes that might classify the data. One reasonable choice as the best hyperplane is the one that represents the largest separation, or margin, between the two classes. So SVM chooses the hyperplane so that the distance from it to the nearest data point on each side is maximized. If such a hyperplane exists, it is known as the maximum-margin hyperplane, and the defined linear classifier is known as a maximum-margin classifier or, equivalently, the perceptron of optimal stability.

To describe the algorithm of SVM, let us consider a two-dimensional classification problem which is divided by a line as shown in Fig. 6.2. The equation of this line is written as

$$W^{\mathrm{T}}X + b = 0 \tag{6.1}$$

where X is the variables in the decision space and W and b are the parameters of the classifier.

In contrast to the artificial neural networks, SVM considers a margin for a classifier line as shown in Fig. 6.3b. Let us define the equations of the marginal lines as

$$W^{\mathrm{T}}X + b = 1 \tag{6.2}$$

and

$$W^{\mathrm{T}}X + b = -1 \tag{6.3}$$

We know that the distance of a line from the origin is obtained by $|b|/\|W\|$. Therefore, the distance between the upper marginal line and the classifier line of Fig. 6.2 is obtained as

$$d = \left| \frac{|b-1|}{\|W\|} - \frac{|b|}{\|W\|} \right| = \frac{1}{\|W\|} \tag{6.4}$$

So the width of the margin is

$$D = 2d = \frac{2}{\|W\|} \tag{6.5}$$

The objective function of SVM method becomes maximizing $2/\|W\|$ or its equivalent form of

$$\text{Min} \quad L = \frac{1}{2} W^{\mathrm{T}} W \tag{6.6}$$

Subject to the following constraints,

$$\begin{array}{llll} \text{if} & y = 1 & \text{then} & W^{\mathrm{T}}X + b \geq 1 \\ \text{if} & y = -1 & \text{then} & W^{\mathrm{T}}X + b \leq -1 \end{array} \tag{6.7}$$

In the above constraints, y is representative of each class. Two above constraints can be combined to the form of

$$\text{Minimize} \quad L = \frac{1}{2} W^{\mathrm{T}} W \tag{6.8}$$

Subject to

$$y\left(W^{\mathrm{T}}X + b\right) - 1 \geq 0 \tag{6.9}$$

The above optimization problem is used for the approach of "hard margin" where a solid border is considered for the support vectors. We can also make the support vectors more flexible by accepting an error of ξ for each of the border lines. So, the optimization becomes

$$\text{Minimize} \quad \frac{1}{2} W^{\mathrm{T}} W + C \sum_{i=1}^{n} \xi_i \quad i = 1, \ldots, n \tag{6.10}$$

$$\begin{array}{lll} \text{Subject to}: & y_i\left(W^{\mathrm{T}}X + b\right) \geq 1 - \xi_i, & \forall i \in \{1, \ldots, n\} \\ & \xi_i \geq 0, & \forall i \in \{1, \ldots, n\} \end{array} \tag{6.11}$$

where n is the number of observation data.

The objective function of Eq. 6.8 could be changed to the following form inserting the constraints into the objective function:

$$\text{Minimize} \quad L = \frac{1}{2} W^{\mathrm{T}} W - \sum_{i} \alpha_i \left[y\left(W^{\mathrm{T}}X + b\right) - 1 \right] \quad i = 1, \ldots, n \tag{6.12}$$

where α is the multiplier of the constraint when it is inserted in the objective function.

The above equation is the primal problem of SVM method. "b" and "W" can be omitted from the equation taking derivations of L.

$$\begin{cases} \dfrac{dL}{dW} = 0 \Rightarrow W - \sum_i \alpha_i y_i x_i \Rightarrow W = \sum \alpha_i y_i x_i \\ \dfrac{dL}{db} = 0 \Rightarrow \sum_i \alpha_i y_i = 0 \end{cases} \tag{6.13}$$

The dual problem is then obtained as follows, inserting the equivalents of b and W:

$$\text{Maximize} \quad L_D = -\frac{1}{2}\sum_i \sum_j \alpha_i \alpha_j y_i y_j x_i^{\mathrm{T}} x_j + \sum_i \alpha_i \quad i = 1, \ldots, n \tag{6.14}$$

$$\text{Subject to} \quad \sum_i \alpha_i y_i = 0 \quad \alpha_i \geq 0 \tag{6.15}$$

The decision space variables (X) could be mapped to a higher dimensional space using function ϕ. Applying this transformation, the dual problem becomes

$$\begin{aligned} &\text{Maximize} && \sum_{i=1}^{n} \alpha_i - \frac{1}{2}\sum_{i,j=1}^{n} \alpha_i \alpha_j y_i y_j . k(x_i, x_j) \\ &\text{Subject to :} && \alpha_i \geq 0, \quad \forall i \in \{1, \ldots, n\} \\ &&& \sum_{i=1}^{n} \alpha_i y_i = 0 \end{aligned} \tag{6.16}$$

where

$$k(x_i, x_j) = (\phi(x_i).\phi(x_j)) \tag{6.17}$$

is called the *kernel* function.

The kernel functions commonly used in SVM's formulations are:

Linear:

$$k(x_i x_j) = x_i^{\mathrm{T}} x_j \tag{6.18}$$

Polynomial:

$$k(x_i x_j) = (\gamma \ x_i^{\mathrm{T}} x_j + r)^d, \quad \gamma > 0 \tag{6.19}$$

Table 6.1 Data presented for Example 6.1

Class A		Class B	
7	15	9.1	10.6
16.3	13.4	6.7	7.3
9.7	20.5	10.3	11.4
11	17.1	10.6	13.2
17.6	19.5	7.7	8.6
16.1	18.7	9.9	6.8
10.4	17.9	10.7	7.1
14.7	18.4	9.6	9.2
9.9	17.8	11.5	11.4
18.1	13.7	8.8	11.6
17.5	13.1	9.7	12.6
11.8	13.8	10.2	11.3
16.9	15.6	10.1	7.6
14.9	22.2	9.8	10
14.8	13.2	8.3	9.7

Radial basis function (RBF):

$$k(x_i x_j) = \exp\left(-\gamma \|x_i - x_j\|^2\right), \quad \gamma > 0 \tag{6.20}$$

Sigmoid:

$$k(x_i x_j) = \tanh\left(\gamma \; x_i^{\mathrm{T}} x_j + r\right). \tag{6.21}$$

where, γ, r, and d are *kernel* parameters.

Example 6.1: Linear Classification by SVM

Classify the data of Table 6.1.

Solution

```
n=numel(y);

ClassA=find(y==1);
ClassB=find(y==-1);


%% Development of SVM

%% Parameter C
C=10;
```

(continued)

```
H=zeros(n,n);
for i=1:n
for j=i:n
    H(i,j)=y(i)*y(j)*x(:,i)'*x(:,j);
    H(j,i)=H(i,j);
end
end

f=-ones(n,1);

Aeq=y;
beq=0;
lb=zeros(n,1);
ub=C*ones(n,1);
Alg='interior-point-convex';
options=optimset('Algorithm',Alg,...
'Display','off',...
'MaxIter',20);

alpha=quadprog(H,f,[],[],Aeq,beq,lb,ub,[],
options)';

AlmostZero=(abs(alpha)<max(abs(alpha))/1e5);
alpha(AlmostZero)=0;


S=find(alpha>0 & alpha<C);

w=0;
for i=S
  w=w+alpha(i)*y(i)*x(:,i);
end


b=mean(y(S)-w'*x(:,S));

%% Plot

Line=@(x1,x2) w(1)*x1+w(2)*x2+b;
LineA=@(x1,x2) w(1)*x1+w(2)*x2+b+1;
LineB=@(x1,x2) w(1)*x1+w(2)*x2+b-1;
```

(continued)

```
figure;
plot(x(1,ClassA),x(2,ClassA),'ro');
hold on;
plot(x(1,ClassB),x(2,ClassB),'bs');
plot(x(1,S),x(2,S),'ko','MarkerSize',12);
x1min=min(x(1,:));
x1max=max(x(1,:));
x2min=min(x(2,:));
x2max=max(x(2,:));

handle=ezplot(Line,[x1min x1max x2min x2max]);
set(handle,'Color','k','LineWidth',2);

handleA=ezplot(LineA,[x1min x1max x2min x2max]);
set
(handleA,'Color','k','LineWidth',1,'LineStyle',',:');

handleB=ezplot(LineB,[x1min x1max x2min x2max]);
set
(handleB,'Color','k','LineWidth',1,'LineStyle',':');

legend('Class A','Class B');
```

the result is shown in Fig. 6.3.

Example 6.2: Nonlinear Classification by SVM

First we need to save the following function in the address where MATLAB is running. The following commands could be copied to an *m*.file then be saved in the address:

```
function f=MySVRFunc(X,alpha,y,x,Kernel)

  n=numel(alpha);

  f=0;
for i=1:n
    f=f+alpha(i)*y(i)*Kernel(x(:,i),X);
end
```

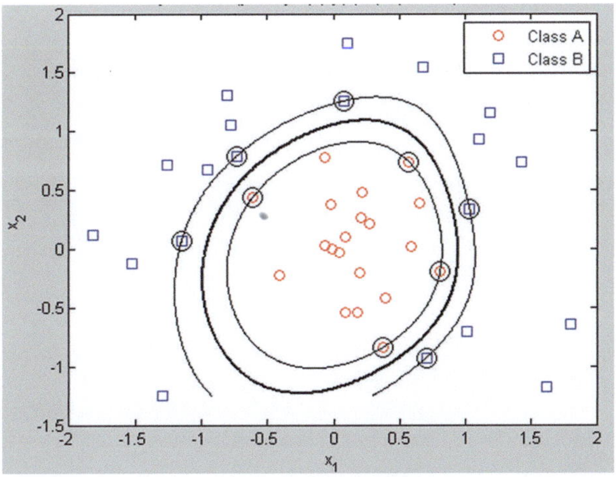

Fig. 6.4 Plot of the classification results in Example 6.2

Then the program is run as follows. The result is shown in Fig. 6.4.

```
%% create data

m=20;
rA=unifrnd(0,1,1,m);
rB=unifrnd(1,2,1,m);
r=[rA rB];
theta=unifrnd(0,2*pi,1,2*m);
x(1,:)=r.*cos(theta);
x(2,:)=r.*sin(theta);
y(1:m)=1;
y(m+1:2*m)=-1;
ClassA=find(y==1);
ClassB=find(y==-1);
n=numel(y);

%% Develop SVM

C=250;

sigma=0.9;

Kernel=@(xi,xj) exp(-1/(2*sigma^2)*norm(xi-xj)^2);
H=zeros(n,n);
for i=1:n
```

(continued)

```
for j=i:n
    H(i,j)=y(i)*y(j)*Kernel(x(:,i),x(:,j));
    H(j,i)=H(i,j);
end
end

f=-ones(n,1);
Aeq=y;
beq=0;
lb=zeros(n,1);
ub=C*ones(n,1);
Alg{1}='trust-region-reflective';
Alg{2}='interior-point-convex';
Alg{3}='active-set';
options=optimset('Algorithm',Alg{2},...
'Display','off',...
'MaxIter',20);

alpha=quadprog(H,f,[],[],Aeq,beq,lb,ub,[],
options)';
AlmostZero=(abs(alpha)<max(abs(alpha))/1e5);
alpha(AlmostZero)=0;
S=find(alpha>0 & alpha<C);
b=0;


b=0;
for i=S
     b=b+y(i)-MySVRFunc(x(:,i),alpha(S),y(S),x(:,S),
Kernel);
end
b=b/numel(S);


%% Plot

Curve=@(x1,x2)MySVRFunc([x1;x2],alpha(S),y(S),x(:,
S),Kernel)+b;
CurveA=@(x1,x2)MySVRFunc([x1;x2],alpha(S),y(S),x(:,
S),Kernel)+b+1;
CurveB=@(x1,x2)MySVRFunc([x1;    x2],alpha(S),y(S),x
(:,S),Kernel)+b-1;

figure;
plot(x(1,ClassA),x(2,ClassA),'ro');
```

(continued)

```
hold on;
plot(x(1,ClassB),x(2,ClassB),'bs');
plot(x(1,S),x(2,S),'ko','MarkerSize',12);
x1min=min(x(1,:));
x1max=max(x(1,:));
x2min=min(x(2,:));
x2max=max(x(2,:));

handle=ezplot(Curve,[x1min x1max x2min x2max]);
set(handle,'Color','k','LineWidth',2);

handleA=ezplot(CurveA,[x1min x1max x2min x2max]);
set
(handleA,'Color','k','LineWidth',1,'LineStyle',':');

handleB=ezplot(CurveB,[x1min x1max x2min x2max]);
set
(handleB,'Color','k','LineWidth',1,'LineStyle',':');
legend('Class A','Class B');
```

When one deals with more than two classes, an appropriate multiclass method is needed. A number of possible methods for this purpose are as follows (Chapelle et al. 1999):

- Modifying the design of the SVM to incorporate the multiclass learning directly in the quadratic solving algorithm
- Combining several binary classifiers with two methods:
  - "One against one" which applies pair comparisons between classes
  - "One against the others" which compares a given class with all the other classes

According to a comparison study (Weston and Watkins 1998), the accuracy of these methods is almost the same.

6.3 Support Vector Machines for Regression

The basic difference between the application of SVM for regression (SVR) and the application of SVM for classification is that in SVR y is considered as a real number instead of a binary number. It means that

$W^{\mathrm{T}}X + b$ in Eqs. 6.1, 6.2, and 6.3 is considered equal to Y or

$$W^{\mathrm{T}}X + b = Y \tag{6.22}$$

where

$$Y \in R$$

Here the goal of modeling will be obtaining the parameters of Eq. 6.14 in a way that the estimated Y be approximately equal to the target values of T for n observation data, when an error less than ε ($\varepsilon = |Y - T|$) is allowed and more than that gets penalty in the objective function.

Like the nonlinear SVM, a mapping of X could be used in Eq. 6.22, which enables using the presented kernel function through the SVR to improve it as a nonlinear regression method.

$$W^{\mathrm{T}}\phi(X) + b = Y \tag{6.23}$$

Example 6.3: Linear Regression Modeling by SVR

Develop a linear regression model by SVR.

Solution

The following program fits a linear regression model to the x and t data as specified in the program:

```
x=linspace(0,10,40);
t=-2*x+7+1*randn(size(x));
n=numel(t);

%% Develop SVR

% epsilon
epsilon=1;
C=1;
H=zeros(n,n);
for i=1:n
```

(continued)

```
for j=i:n
    H(i,j)=x(:,i)'*x(:,j);
    H(j,i)=H(i,j);
end
end

% HH=H
HH=[ H -H
  -H H];

f=[-t'; t']+epsilon;


Aeq=[ones(1,n) -ones(1,n)];
beq=0;
lb=zeros(2*n,1);
ub=C*ones(2*n,1);
options=optimset('Display','iter','MaxIter',1000);
alpha=quadprog(HH,f,[],[],Aeq,beq,lb,ub,[],
options);
alpha=alpha';
AlmostZero=(abs(alpha)<max(abs(alpha))*1e-4);
alpha(AlmostZero)=0;
alpha_plus=alpha(1:n);
alpha_minus=alpha(n+1:end);


eta=alpha_plus-alpha_minus;

S=find(alpha_plus+alpha_minus>0
  &alpha_plus+alpha_minus<C);


w=0;
for i=1:n
  w=w+eta(i)*x(:,i);
end

b=mean(t(S)-w'*x(:,S)-sign(eta(S))*epsilon);

%% Plot
```

(continued)

Fig. 6.5 Plot of the regression line in Example 6.3

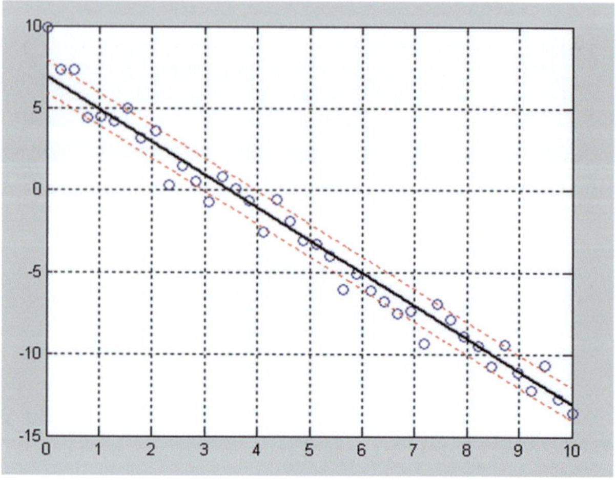


```
figure;
plot(x,t,'o');
hold on;
xmin=min(x);
xmax=max(x);
xx=linspace(xmin,xmax,500);
yy=w'*xx+b;
plot(xx,yy,'k','LineWidth',2);
plot(xx,yy+epsilon,'r:');
plot(xx,yy-epsilon,'r:');
grid on;
```

The result is shown in Fig. 6.5.

Example 6.4: Nonlinear Regression Modeling by SVR

Develop a nonlinear regression by SVR using a kernel function.

Solution
First we need to save the following function in the address where MATLAB is running. The following commands could be copied to an *m*.file then be saved in that address:

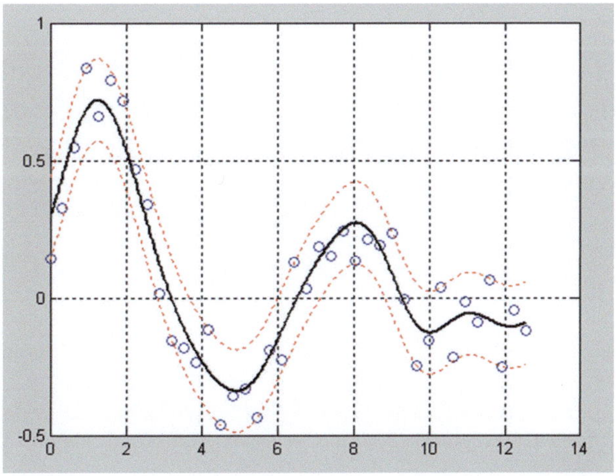

Fig. 6.6 Plot of the regression curve in Example 6.4

```
function y=MySVRFunc(x,eta,X,Kernel)

  n=numel(eta);
  y=0;
for i=1:n
    y=y+eta(i)*Kernel(X(:,i),x);
end

end
```

Then the following program is run. The output is seen in Fig. 6.6.

```
x=linspace(0,4*pi,40);
t=sin(x).*exp(-0.2*x)+0.1*randn(size(x));
n=numel(t);


%% Design SVR

epsilon=0.15;
C=1000;
sigma=1;

Kernel=@(xi,xj) exp(-1/(2*sigma^2)*norm(xi-xj)^2);

H=zeros(n,n);
for i=1:n
```

(continued)

```
for j=i:n
    H(i,j)=Kernel(x(:,i),x(:,j));
    H(j,i)=H(i,j);
end
end

HH=[ H -H
  -H H];

f=[-t'; t']+epsilon;

Aeq=[ones(1,n) -ones(1,n)];
beq=0;

lb=zeros(2*n,1);
ub=C*ones(2*n,1);

options=optimset('Display','iter','MaxIter',1000);

alpha=quadprog(HH,f,[],[],Aeq,beq,lb,ub,[],
options);

alpha=alpha';

AlmostZero=(abs(alpha)<max(abs(alpha))*1e-4);

alpha(AlmostZero)=0;

alpha_plus=alpha(1:n);
alpha_minus=alpha(n+1:end);

eta=alpha_plus-alpha_minus;

S=find(alpha_plus+alpha_minus>0
  &alpha_plus+alpha_minus<C);

y=zeros(size(t));
for i=1:n
  y(i)=MySVRFunc(x(:,i),eta(S),x(:,S),Kernel);
end

b=mean(t(S)-y(S)-sign(eta(S))*epsilon);
```

(continued)

```
%% Plot

xmin=min(x);
xmax=max(x);
xx=linspace(xmin,xmax,500);
yy=zeros(size(xx));
for k=1:numel(yy)
yy(k)=MySVRFunc(xx(:,k),eta(S),x(:,S),Kernel)+b;
end

figure;
plot(x,t,'o');
hold on;
plot(xx,yy,'k','LineWidth',2);
plot(xx,yy+epsilon,'r:');
plot(xx,yy-epsilon,'r:');
grid on;
```

References

Burges CJC (1998) A tutorial on support vector machines for pattern recognition, vol 2. Kluwer Academic, Boston, pp 1–43

Chapelle O, Haffner P, Vapnik VN (1999) Support vector machines for histogram-based classification. IEEE Trans Neural Netw 10(5):1055–1064

Vapnik VN (1995) The nature of statistical learning theory. Springer, New York. ISBN 0-387-98780-0

Weston J, Watkins C (1998) Multiclass support vector machines. Technical report. CSD-TR-98-04, Department of Computer Science, Royal Holloway, University of London, Egham

Chapter 7
Fuzzy Models

Abstract While variables in mathematics usually take numerical values, in fuzzy logic applications, the non-numeric linguistic variables are often used to facilitate the expression of rules and facts. The idea of fuzzy logic is very suitable for engineering application where a precise representation of the real world is sought. In contrast to the statistical-based methods, fuzzy models do not need very strong assumptions and requirements. As far as the engineering application of fuzzy logic is concerned, two approaches are usually followed up: (1) developing fuzzy extensions of the classic methods and models and (2) developing models, which are basically originated by the fuzzy logic. Basic information in fuzzy logic, fuzzy clustering, fuzzy inference systems, and fuzzy regression are the main subjects which are presented in this chapter. Obviously, the related useful MATLAB commands are presented and discussed to support the methods of applied modeling of the presented subjects.

Keywords Fuzzy logic • Fuzzy C-means • Fuzzy inference systems • Adaptive neuro-fuzzy inference system • Fuzzy regression

7.1 Introduction

Application of fuzzy logic in the engineering fields has increased significantly in recent years. While variables in mathematics usually take numerical values, in fuzzy logic applications, the non-numeric linguistic variables are often used to facilitate the expression of rules and facts. A linguistic variable such as *quality* may have a value such as *appropriate* or its antonym *inappropriate*. However, the great utility of linguistic variables is that they can be modified via linguistic hedges applied to primary terms. This general idea is very suitable for engineering applications where a precise representation of the real world is sought. A summary review of the application of fuzzy logic in water resources and environmental engineering contests is shown in Table 7.1.

S. Araghinejad, *Data-Driven Modeling: Using MATLAB® in Water Resources and Environmental Engineering*, Water Science and Technology Library 67,
DOI 10.1007/978-94-007-7506-0_7,

Table 7.1 A summary review on the application of fuzzy logic in water resources and environmental engineering

Researcher(s)	Field of the study	Subject
Ren et al. (2013)	Hydrology	Monthly runoff forecasting with adaptive neuro-fuzzy inference system and wavelet analysis
Wang and Altunkaynak (2012)	Hydrology	A comparative assessment of rainfall–runoff modeling between SWMM and fuzzy logic approach
Kucukmehmetoglu et al. (2010)	Water resources management	Coalition possibility of riparian countries via game theory and fuzzy logic models
Shrestha and Simonovic (2010)	Hydrometry	Fuzzy nonlinear regression approach to stage–discharge analyses
Talei et al. (2010)	Hydrology	Evaluation of rainfall and discharge inputs used by adaptive network-based fuzzy inference systems (ANFIS) in rainfall–runoff modeling
Firat et al. (2009)	Water consumption prediction	Comparative analysis of fuzzy inference systems for prediction of water consumption time series
Lee et al. (2009)	Urban water systems	Fuzzy logic modeling of risk assessment for a drinking-water supply system
Moghaddamnia et al. (2009)	Hydroclimatology	Evaporation estimation, using artificial neural networks and adaptive neuro-fuzzy inference system techniques
Mathon et al. (2008)	Groundwater	Transmissivity and storage coefficient estimation by coupling the Cooper–Jacob method and modified fuzzy least-squares regression
Shu and Ouarda (2008)	Hydrology	Regional flood frequency analysis at un-gauged sites using the adaptive neuro-fuzzy inference system
Casper et al. (2007)	Hydrology	Fuzzy logic-based rainfall–runoff modeling using soil moisture measurements
Lohani et al. (2007)	Sediment estimation	Deriving stage–discharge–sediment concentration relationships using fuzzy logic
Chang and Chang (2006)	Reservoir management	Adaptive neuro-fuzzy inference system for prediction of water level in reservoirs
Terzi et al. (2006)	Climatology	Estimation of evaporation using ANFIS
Özelkan et al. (1996)	Climatology	Relationship between monthly atmospheric circulation patterns and precipitation
Bardossy et al. (1990)	Hydrology	Application of fuzzy regression in hydrology

As far as the engineering application of fuzzy logic is concerned, two types of fuzzy models exist: (1) models that are in fact fuzzy extensions of the classic ones and (2) models which are basically originated by the fuzzy logic. Fuzzy regression is a model which is categorized in the first type of the mentioned models, and fuzzy inference system is among the models of type two.

More generally, fuzzy logic, neuro-computing, and genetic algorithms may be viewed as the principal constituents of what might be called soft computing. Unlike the traditional, hard computing, soft computing accommodates the imprecision of

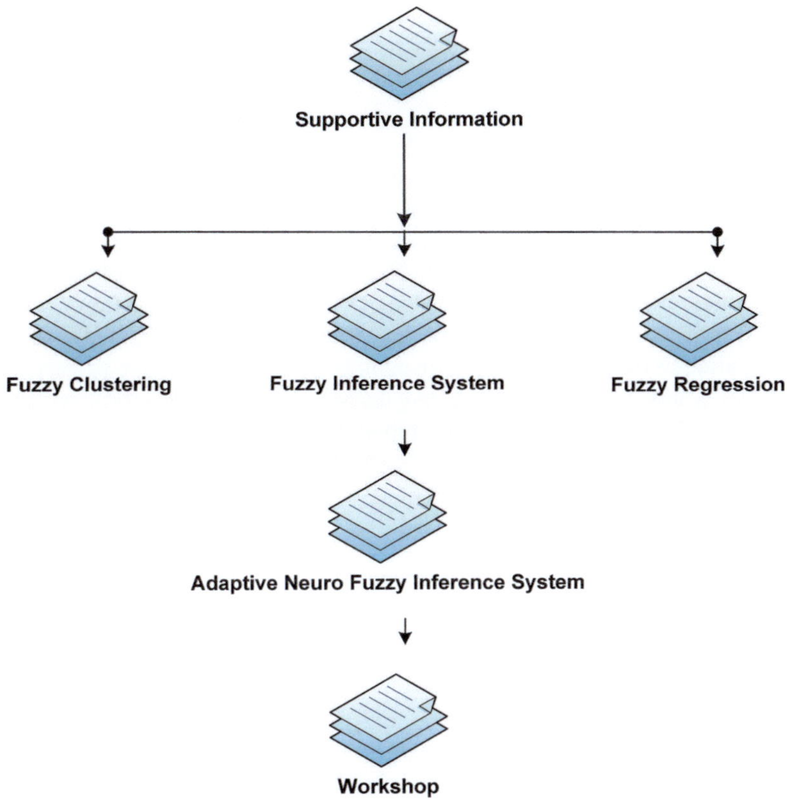

Fig. 7.1 The structure of contents of this chapter

the real world. Among various combinations of methodologies in soft computing, the one that has highest visibility at this juncture is that of fuzzy logic and neuro-computing, leading to neuro-fuzzy systems. Within fuzzy logic, such systems play a particularly important role in the induction of rules from observations.

This chapter is structured as shown in the chart of Fig. 7.1. After "*Introduction*," some supportive information regarding the fuzzy sets is presented. The next section deals with one of the application of fuzzy logics in the field of data clustering. The well-known technique of fuzzy C-means is presented in this chapter. The chapter is continued by the most well-known data-driven fuzzy model, fuzzy inference system. Two techniques of *Mamdani* and *Sugeno* fuzzy inference systems are presented in that section. An improved technique of fuzzy inference system by neuro-computing, which is known as adaptive neuro-fuzzy inference system, is presented in the next section. The final section of this chapter deals with a fuzzy version of linear regression. Finally, the chapter ends with a workshop on the fuzzy systems.

7.2 Supportive Information

7.2.1 *Fuzzy Numbers*

Fuzzy logic starts with the concept of a fuzzy set. A fuzzy set is a set without a crisp, clearly defined boundary. To understand what a fuzzy set is, first consider what is meant by what we might call a classic set. A classic set usually deals with deterministic and random variables as shown in Fig. 7.2. The presentation of variables in a classic manner includes either a crisp value or the range of probable values with their associated probabilities. A vast range of mathematical and statistical methods exist to describe variables in a classic manner. In contrast, fuzzy sets describe vague concepts (severity of a drought, hot weather, and benefit of water allocation). A fuzzy set admits the possibility of partial membership in it (quality of water is poor, the weather

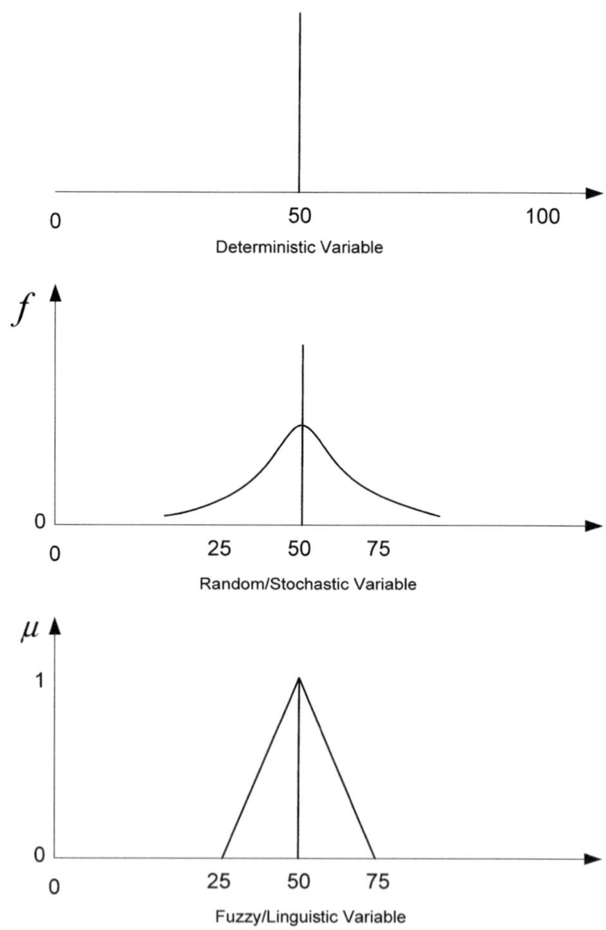

Fig. 7.2 Variables in mathematics, statistics, and fuzzy logic

Table 7.2 Conventional fuzzy membership function in fuzzy logic application

Type of fuzzy membership functions (MF)	Syntax of the MFs in MATLAB
Triangular	Trimf
Trapezoidal	Trapmf
Gaussian	Guassmf
Generalized bell	Guass2mf
Sigmoidal	Sigmf

is rather hot). A fuzzy number is an extension of a regular number in the sense that it does not refer to one single value but rather to a connected set of possible values, where each possible value has its own weight between 0 and 1. This weight is called the membership function.

The membership function is the basic idea in fuzzy set theory; its values measure degrees to which objects satisfy imprecisely defined properties.

Fuzzy memberships represent similarities of objects to imprecisely defined properties. Membership values determine how much fuzziness a fuzzy set contains.

The degree an object, x, belongs to a fuzzy set; $\widetilde{A}$ is denoted by a membership value between 0 and 1 known as $\mu\widetilde{A}(x)$. A membership function associated with a given fuzzy set maps an input value to its appropriate membership value. Types of conventional membership functions are listed in Table 7.2.

For instance, a triangular membership function with a and c as the feet of the shape and b as the peak is developed by

```
y= trimf(x,[a b c])
```

where $x =$ input vector and $y = \mu_{\widetilde{A}}(x) =$ output vector in the range [0,1]. $\mu_{\widetilde{A}}(x)$ is related to the x values by the following relation:

$$\mu_{\widetilde{A}}(x) = \begin{cases} 0 & \text{if} \quad x < a \\ \dfrac{x-a}{b-a} & \text{if} \quad a \le x \le b \\ \dfrac{c-x}{c-b} & \text{if} \quad b \le x \le c \\ 0 & \text{if} \quad x > c \end{cases} \tag{7.1}$$

Figure 7.3 shows a triangular fuzzy number obtained by the following commands in MATLAB.

```
x=0:1:20;
y=trimf(x,[5 10 18]);
plot(x,y)
xlabel('trimf, P=[5 10 18]')
```

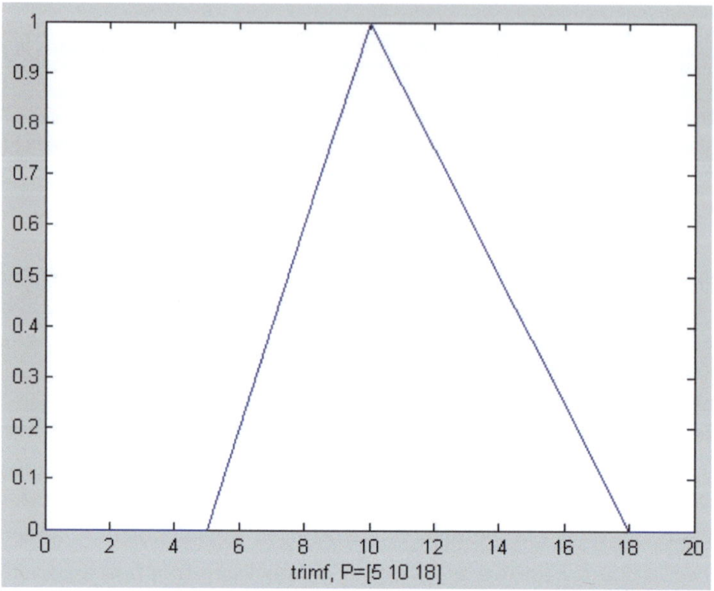

Fig. 7.3 An example of triangular fuzzy membership function

In a trapezoidal membership function, the membership value, $\mu_{\tilde{A}}(x)$, is obtained as

$$\mu_{\tilde{A}}(x) = \begin{cases} 0 & \text{if} \quad x < a \\ \dfrac{x-a}{b-a} & \text{if} \quad a \le x \le b \\ 1 & \text{if} \quad b \le x \le c \\ \dfrac{d-x}{d-c} & \text{if} \quad c \le x \le d \\ 0 & \text{if} \quad x > d \end{cases} \tag{7.2}$$

Figure 7.4 shows a trapezoidal membership function obtained by the following commands:

```
x=0:1:20;
y=trapmf(x,[5 10 15 18]);
plot(x,y)
xlabel('trapmf, P=[5 10 15 18]')
```

The membership value, $\mu_{\tilde{A}}(x)$, by a Gaussian membership function is obtained by

$$\mu_{\tilde{A}}(x) = e^{\frac{-(x-c)^2}{2\sigma^2}} \tag{7.3}$$

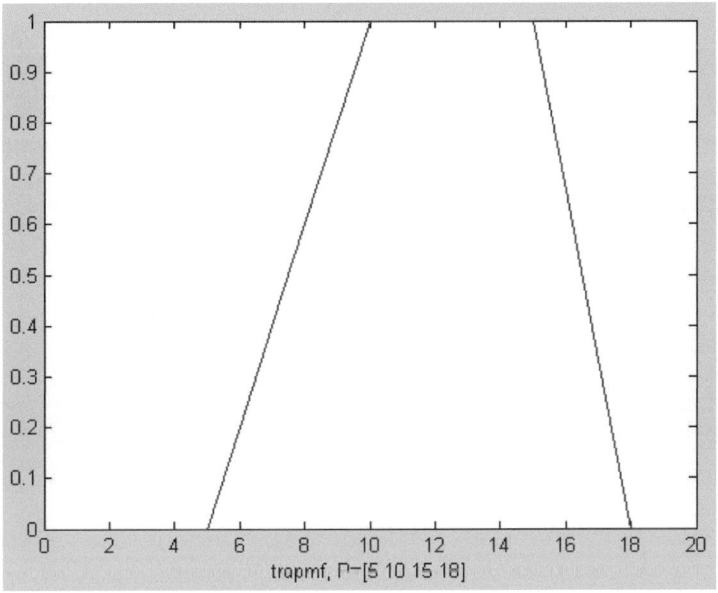

Fig. 7.4 An example of trapezoidal fuzzy membership function

An example of Gaussian membership function is presented in Fig. 7.5 which has been obtained by the following commands:

```
x=0:1:20;
sigma=2;
average=10;
y=gaussmf(x,[sigma average]);
plot(x,y)
xlabel('gaussmf, P=[2 10]')
```

7.2.2 Logical Operators

To combine two fuzzy numbers, AND, OR, and NOT operators exist in fuzzy logic, usually defined as the minimum, maximum, and complement. It means that the membership function of the output variable in case of combining two fuzzy numbers would be the maximum/minimum membership function of them in case of using AND/OR operators.

In case of using NOT operator, the membership function of the output variable would be the complement of the single input variable. Table 7.3 shows an example of combining two fuzzy variables by different operators.

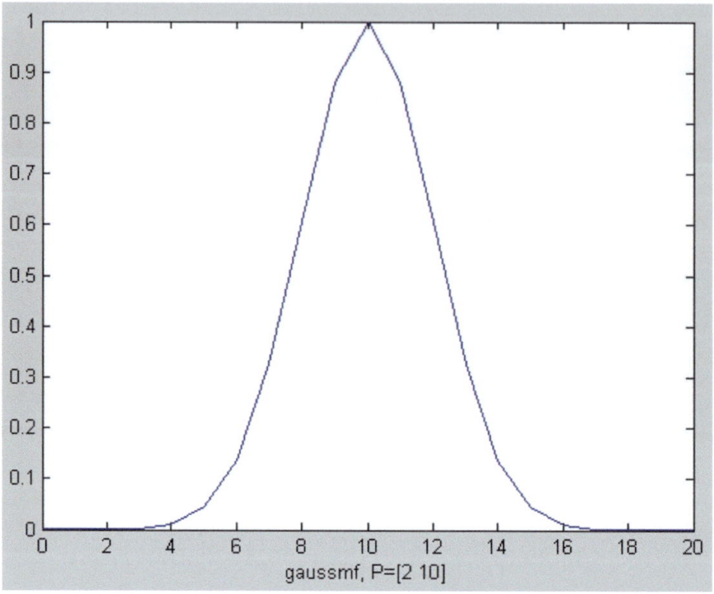

Fig. 7.5 An example of Gaussian fuzzy membership function

Table 7.3 Operators in fuzzy logic

	Membership values		Result
Operator	$\mu_{\tilde{A}}(x) = 0.8$	$\mu_{\tilde{B}}(x) = 0.3$	$\mu_{\tilde{C}}(x)$
A AND *B*	Min(0.8,0.3)		0.3
A OR *B*	Max(0.8,0.3)		0.8
NOT *A*	1–0.8		0.2

7.3 Fuzzy Clustering

Clustering of numerical data forms the basis of many classification and system modeling algorithms. The purpose of clustering is to identify natural groupings of data from a large data set to produce a concise representation of a system's behavior. Clustering is a mathematical tool that attempts to discover structures or certain patterns in a data set, where the objects inside each cluster show a certain degree of similarity.

Hard clustering assigns each feature vector to one. Only one of the clusters with a degree of membership is equal to one, and well-defined boundaries exist between clusters. Fuzzy clustering allows each feature vector to belong to more than one cluster with different membership degrees (between 0 and 1) and vague or fuzzy boundaries between clusters. Fuzzy C-means (FCM) is a data clustering technique wherein each data point belongs to a cluster to some degree that is specified by a membership grade.

For x_i data of size N and C number of clusters, considering c_j as the center of each cluster and u_{ij} as the coefficients of each x_i for being in the clusters, the optimum values of c_j is obtained when the following objective function, J, is minimized:

$$J_m = \sum_{i=1}^{N} \sum_{j=1}^{C} u_{ij}^m \left\| x_i - c_j \right\|^2 \tag{7.4}$$

where m is a parameter to control the weight which is given to each center of clusters.

To minimize the above function, the algorithmic procedure for applying fuzzy C-means for data of size N presented by Bezdek et al. (1984) is as follows:

1. Choose C as the number of clusters ($2 \leq C \leq N$).
2. Select parameter m and assign randomly to each point coefficients for being in the clusters, $u_{ij}^m (i = 1, \ldots, N;\ j = 1, \ldots, C)$. It should be noted that $\sum_{j=1}^{C} u_{ij}^m = 1$. m controls how much weight is given to the closest center.
3. Determine the coordinate of each cluster by the following formula:

$$c_j = \frac{\sum_{i=1}^{N} u_{ij}^m x_i}{\sum_{i=1}^{N} u_{ij}^m} \tag{7.5}$$

4. Update u_{ij}^m values by the following relation:

$$c_j = \frac{\sum_{i=1}^{N} u_{ij}^m x_i}{\sum_{i=1}^{N} u_{ij}^m} \tag{7.6}$$

5. Considering ε as a small number, if $|\text{new } u_{ij}^m - u_{ij}^m| \leq \varepsilon$, stop the procedure and report c_j values as the center of clusters. Otherwise repeat the procedure from step 3, considering $u_{ij}^m = \text{new } u_{ij}^m$.

In MATLAB, the following syntax is used to cluster a "data" set for a specific number of clusters C:

```
[center,U,objFcn] = fcm(data,C);
```

The output of the syntax would be the center of each C clusters, fuzzy partition matrix for each individual data, U, and finally the value of the objective function at the different iterations of the algorithm, $objFcn$.

Example 7.1: Illustrative Example of Fuzzy C-Means Clustering

Investigate the center of two clusters for the following data (Table 7.4).

Table 7.4 Data presented for Example 7.1

	Data		Data
1	17.89	10	18.58
2	24.26	11	16.56
3	38.91	12	14.11
4	18.67	13	13.42
5	20.67	14	13.06
6	23.11	15	12.62
7	35.55	16	16.69
8	32.14	17	17.63
9	22.53	18	26.43

Solution

Using the following syntax:

```
[center,U,objFcn] = fcm(data,2);
```

For two clusters the results would be

```
center =

    17.4868
    33.7956
```

and the partition matrix is given in matrix U. For example, the membership of 24.26 (the second number) to the first cluster is 0.66 where it is 0.33 as far as the second cluster is concerned.

For three clusters we will have the following centers of clusters:

```
center =

   15.6312
   23.2027
   35.7634
```

Example 7.2: Clustering of the Meteorological Data

Table 7.5 demonstrates rainfall and air temperature recorded in a meteorological station. How the data could be divided if two specific clusters are considered to cluster them.

Table 7.5 Data presented for Example 7.2

	Total rainfall	Average air temperature
1	654	5
2	651	6.5
3	745	4.7
4	763	4.2
5	451	9.2
6	467	10.1
7	678	8.6
8	921	4.2
9	522	10.3
10	491	9.9
11	743	7.6
12	608	9.7
13	780	8.7
14	452	9.9

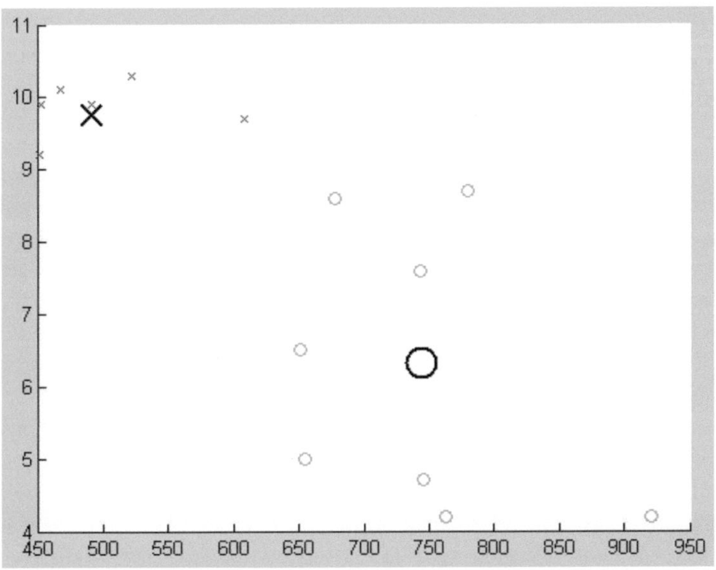

Fig. 7.6 Data clusters in Example 7.2

Solution
The following commands find the centers of two clusters and plot the data according to the cluster they belong:

```
[center, U, objfcn] = fcm(data, 2);
maxU = max(U);
index1 = find(U(1, :) == maxU);
index2 = find(U(2, :) == maxU);
line(data(index1,1), data(index1, 2), 'linestyle',...
```

(continued)

```
'none','marker', 'o','color','g');
line(data(index2,1),data(index2,2),'linestyle',...
'none','marker', 'x','color','r');
hold on
plot(center(1,1),center(1,2),'ko', ...
'markersize',15,'LineWidth',2)
plot(center(2,1),center(2,2),'kx', ...
'markersize',15,'LineWidth',2)
```

As reported by center matrix, the coordinates of the two clusters are (744.2, 6.3) and (490.8, 9.8). The data are plotted in Fig. 7.6.

7.4 Fuzzy Inference System

Fuzzy inference system (FIS) is usually used for mapping a set of input/output data in which there is no predetermined structure. FIS is presented as logical rules named as *if–then* rules. The possibility logic of fuzzy systems is a useful tool for considering the vagueness in modeling real-world problems. What makes the use of FIS more efficient than regression-based methods in some cases is to deal with the vagueness in predictors and the nonlinear relationship between the predictors and dependent variables. Furthermore, FIS can deal with both linguistic and quantitative variables in the process of modeling. In contrast to the conventional data-driven methods, which try to find a logical relationship between input and output variables from the observed data, FIS gets benefit from both the concept of the problem and the information within the observed data.

A fuzzy inference system is based on logical *if–then* rules. A fuzzy *if–then* rule assumes the following form:

If x is in A then y is in B

where A and B are linguistic values defined by fuzzy sets on the ranges X and Y, respectively. The "*if*" part of the rule (x is in A) is called the antecedent or premise, while the "*then*" part of the rule (y is in B) is called the consequence or conclusion. In general, the input to an if–then rule is a single value where the output is a fuzzy set. This set could be defuzzified, assigning one value to the output set.

The following algorithmic steps define how to develop a fuzzy inference system for a specific problem with known input and output variables:

1. Classify each variable to the desired classes getting benefit from the fuzzy numbers. For instance a rainfall variable in the range of 0–100 mm could be represented as three classes of "low rainfall," "medium rainfall," and "high

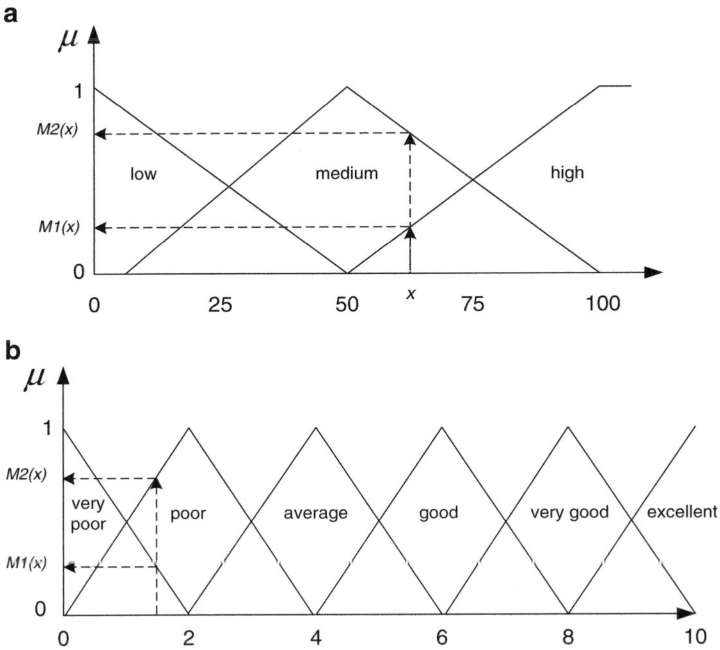

Fig. 7.7 Examples of presenting different variables by fuzzy sets (**a**) Three membership functions (**b**) Six membership functions

rainfall" as presented in Fig. 7.7a. Another example is to classify the quality of water resources for agricultural demand from the minimum 0 to the maximum 10 by the linguistic classes of very poor, poor, average, good, very good, and excellent as shown in Fig. 7.7b. As shown in Fig. 7.7, every variable x could have at least one membership value, $m(x)$, between 0 and 1, which demonstrates how that variable belongs to a specific class. Defining these fuzzy numbers helps developing *if–then* rules in the next step.

2. Define *if–then* rules based on the concept of the problem by the use of defined fuzzy numbers in the previous step. Examples of fuzzy rules could be presented as follows:

   - If rainfall is low and air temperature is high, then a dry season occurs.
   - If total dissolved solid (TDS) of the river is low and water temperature is medium, then water quality is good for environmental needs.
   - If air humidity is high and air temperature is low, then evapotranspiration is low.

   In a fuzzy inference system, the rules are derived based on the investigation of a joint between input and output data. The fuzzy rules are derived based on the frequency of the observation of variables in input and output data.

3. Combine if–then rules to complete the system to be used for modeling. Two common methods of combining the rules are the methods of *Mamdani* and *Sugeno* which are defined as follows:

   - Mamdani inference system

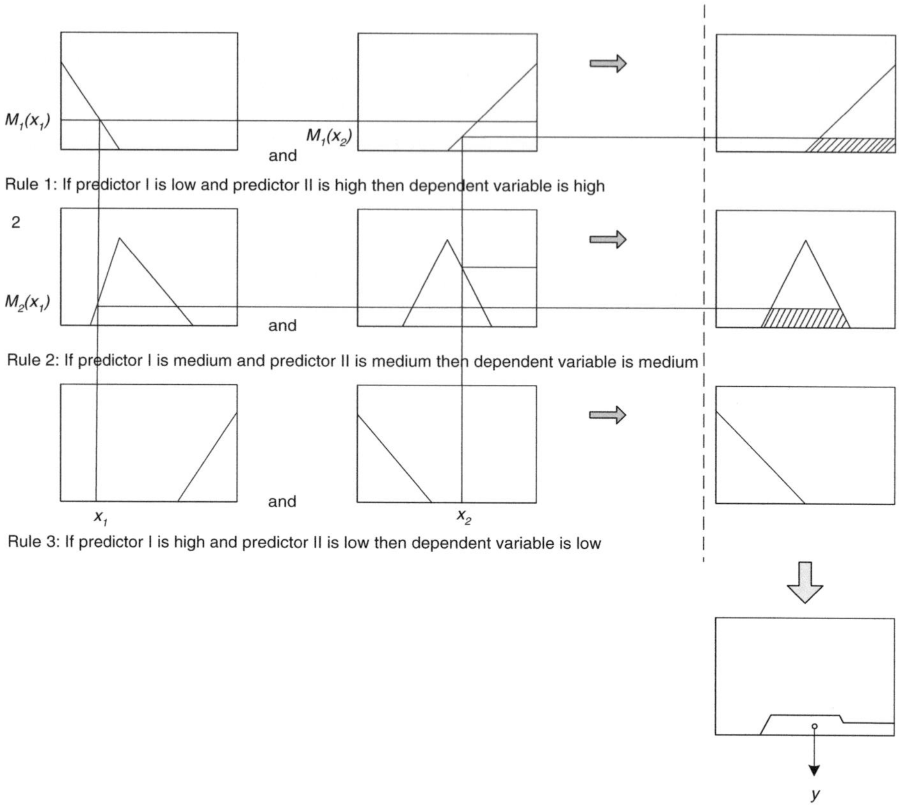

Fig. 7.8 Example of *Mamdani* inference system

Let us assume $x = \{x_1, x_2, \ldots, x_K\}$ as a vector of independent crisp values and the fuzzy response of those values, y, is to be determined using fuzzy rules. The degree of fulfillment of x_k in the membership function of kth independent value in rule i is defined as $m(x_k)$. The degree of fulfillment of the rule r is defined as the minimum among the degrees of fulfillment of all inputs into the rule if "AND" operator is applied to relate the independent values. It is also the maximum among the degrees of fulfillment of all inputs into the rule if "OR" operator is applied. The rule combination in *Mamdani's* method is obtained by superposition of the outputs of each rule. A graphical example of FIS using *Mamdani's* method is shown in Fig. 7.8.

The FIS shown in Fig. 7.8 consisted of three *if–then* rules. Two independent variables (predictors) are considered in this system in order to estimate the predicted value. For the simplicity, all membership functions are considered as triangles. As shown in that figure, the observed values of x_1 for the first predictor and x_2 for the second predictor activate rule 1 and rule 2. Each rule provides a fuzzy set as the output variable, related to the minimum degree of fulfillment of the predictors. The contribution of the output of each rule is related to the minimum values of the degree of fulfillment of the predictors of that rule when "AND" operator is used. The final answer is an aggregation of these outputs, which represent a fuzzy variable and could be defuzzified to represent the representative value of the estimated range of the dependent variable. Defuzzification is the process of producing a quantifiable result in fuzzy logic, given fuzzy sets and corresponding membership degrees. A common defuzzification technique is the center of gravity. First, the results of the rules must be added together. Now, if this triangle were to be cut in a straight horizontal line somewhere between the top and the bottom, and the top portion were to be removed, the remaining portion forms a trapezoid. All of these trapezoids are then superimposed one upon another, forming a single geometric shape. Then, the centroid of this shape, called the fuzzy centroid, is calculated. The coordinate of the centroid is the defuzzified value.

- Sugeno inference system

The *Sugeno* inference system is similar to the *Mamdani's* in terms of developing the fuzzy rules and calculating membership associated with each rule according to the specific input variables and the "AND/OR" operators. Instead of aggregating the fuzzy outputs of different rules, *Sugeno* considers a linear regression model for each rule and aggregates the output of each model by weighted averaging. For each fuzzy rule, i, an output in form of $y_i = \sum_j^m a_j x_j + a_0$ is calculated when m is the number of inputs and a's are the linear regression parameters. In case of a FIS with N fuzzy rules, the final output will be

$$y = \frac{\sum_{i=1}^{N} M_i(x) y_i}{\sum_{i=1}^{N} M_i(x)} \tag{7.7}$$

where $M_i(x)$ is the calculated membership value associated with the ith fuzzy rule. Figure 7.9 shows a *Sugeno* FIS for two inputs, one output and three fuzzy rules.

4. Calibrate the system.

   The location of the center of gravity and the range defining the membership functions of the fuzzy input/output variables in each rule affect the results of modeling. The best value of these characteristics must be determined for each rule after setting the shape of membership function. Determination of the two mentioned characteristics in fuzzy rules is initially set by a first estimation on engineering judgment and then is calibrated to satisfy the following criteria:

Minimum bias: It means that the rules are set to generate minimum point estimation error. Bias is measured as $|y - \hat{y}|$ where y is the actual and $\hat{y}$ is the estimated value.

Minimum spread: It implies that the output fuzzy numbers of estimated value must have the minimum range that is possible. Range of forecasted values is defined as $\hat{y}_{\max} - \hat{y}_{\min}$ where $\hat{y}_{\max}$ is the upper bound and $\hat{y}_{\min}$ is the lower bound of the fuzzy estimated variable.

Maximum reliability: It implies the number of estimations where $\hat{y}_{\min} < y < \hat{y}_{\max}$ to the total number of the estimation values.

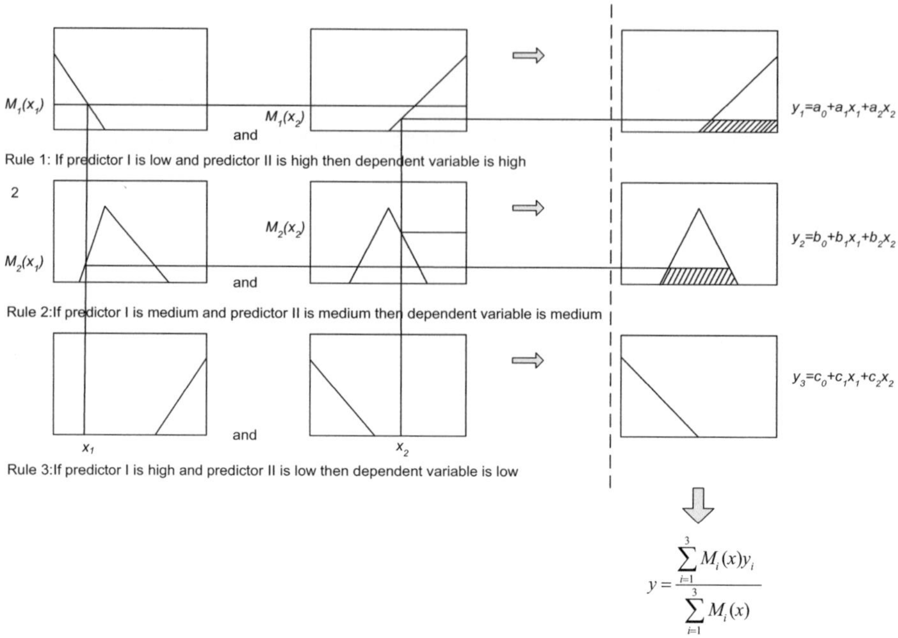

Fig. 7.9 Example of *Sugeno* inference system

Example 7.3: Estimation of Urban Water Consumption Per Capita

A survey has demonstrated that the consumption per capita in a specific month in an urban region depends not only on the weather but also on the occurrence of holidays. As a matter of fact, weather and the type of the day (in terms of being considered as a weekday or a holiday) are considered as predictors of per capita use of urban water. The recoded data shows that the range of water consumption value in that month initiates from 177 l per capita per day and reaches 285 l per capita per day. This is while the averaged air temperature during the month varies from 19° to 25° of centigrade. The survey results in the following *if–then* rules to predict daily urban consumption in a specific day of the month in terms of the introduced predictors and dependent variable:

- If it is *not* a *holiday* and *weather* is *cool,* then *water consumption* per capita is *normal.*
- If it is *not* a *holiday* and *weather* is *warm,* then *water consumption* per capita is *high.*
- If it is a *holiday* and *weather* is *cool,* then *water consumption* per capita is *low.*
- If it is a *holiday* and *weather* is *warm,* then *water consumption* per capita is *normal.*

Table 7.6 Quantitative records for Example 7.3

Holiday	Weather	Water consumption
Long holiday	25	283
holiday	19	177
Week day	19	200

Table 7.7 Suggested fuzzy numbers for the weekdays

Category	Range	Indicator
Not a holiday	0–1	0
Holiday	0–2	1
Long holiday	1–2	2

Table 7.8 Suggested fuzzy numbers for air temperature

Category	Range (°C)	Indicator (°C)
Cool	16–23	19
Warm	20–29	25

Table 7.9 Suggested fuzzy numbers for water consumption per capita

Category	Range (liter per day per person)	Indicator (liter per day per person)
Low	160–200	170
Normal	180–220	200
High	200–260	230
Very high	260–330	280

- If it is a *long holiday* and *weather* is *cool,* then *water consumption* per capita is *high.*
- If it is a *long holiday* and *weather* is *warm,* then *water consumption* per capita is *very high.*

Table 7.6 shows some quantitative records that could help calibrating the above linguistic rules.

Develop a fuzzy inference system to model the urban water use due to the variation of the predictors in the region.

Solution

1. Predictors and dependent variables are represented in form of fuzzy numbers as suggested in Tables 7.7, 7.8, and 7.9.
2. To develop a fuzzy inference system within MATLAB, type the following syntax in the command window to activate the FIS graphical user interface:

```
Fuzzy
```

A fuzzy interface system is developed by introducing the number of inputs and outputs and a desired method between *Mamdani* and *Sugeno*. Figure 7.10 shows that for the current problem, a map of two to one is considered through the *Mamdani* method:

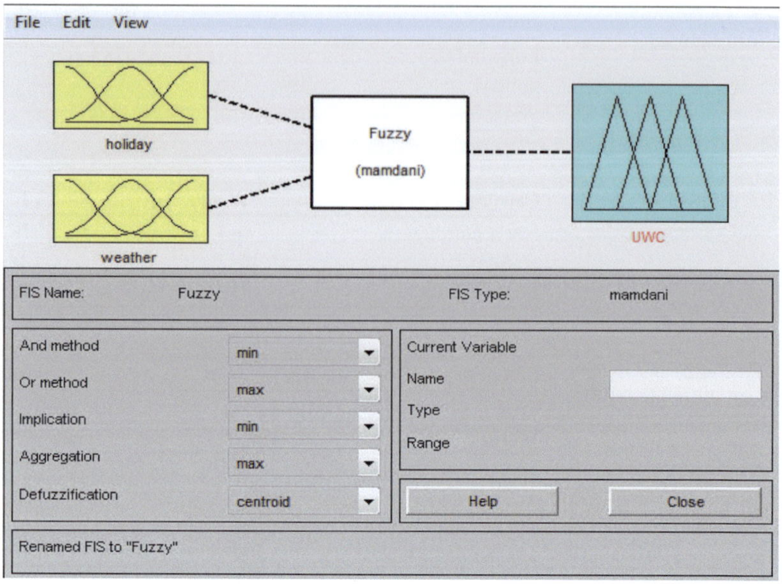

Fig. 7.10 FIS editor in MATLAB

3. Fuzzy numbers are defined within each input and output boxes starting by double clicking on them. For instance, the fuzzy numbers for the first input variable as well as the output variable (urban water consumption, UWC) are shown in the following figures (Figs. 7.11 and 7.12):
4. Develop the predefined if–then rules by the "rule editor" within the "edit" menu (Fig. 7.13).
5. View the *Mamdani* fuzzy inference system based on the defined fuzzy numbers and fuzzy rules within the "*rule viewer*" (Fig. 7.14).
6. The last but not the least step is to validate the system by experimenting different input/output data sets. First of all the estimated output for the test data must be in agreement with the fuzzy rules. Additionally they must provide the same results as recorded by the observed data. In case that there is a deviation between the calculated and observed data, the system could be tuned by changing the range and/or indicators of fuzzy numbers as well as type of the membership functions.

7.5 Adaptive Neuro-Fuzzy Inference System

Adaptive neuro-fuzzy inference system (ANFIS) is a kind of neural network that is based on *Sugeno* fuzzy inference system. Since it integrates both neural networks and fuzzy logic principles, it has potential to capture the benefits of both in a single framework. Its inference system corresponds to a set of fuzzy *if–then rules* that have learning capability to approximate nonlinear functions.

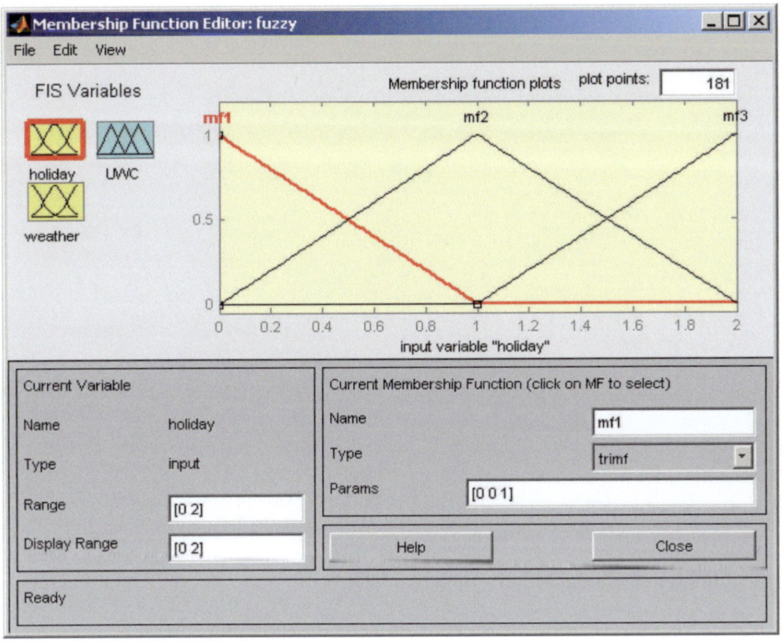

Fig. 7.11 Fuzzy variables for the first input of Example 7.3

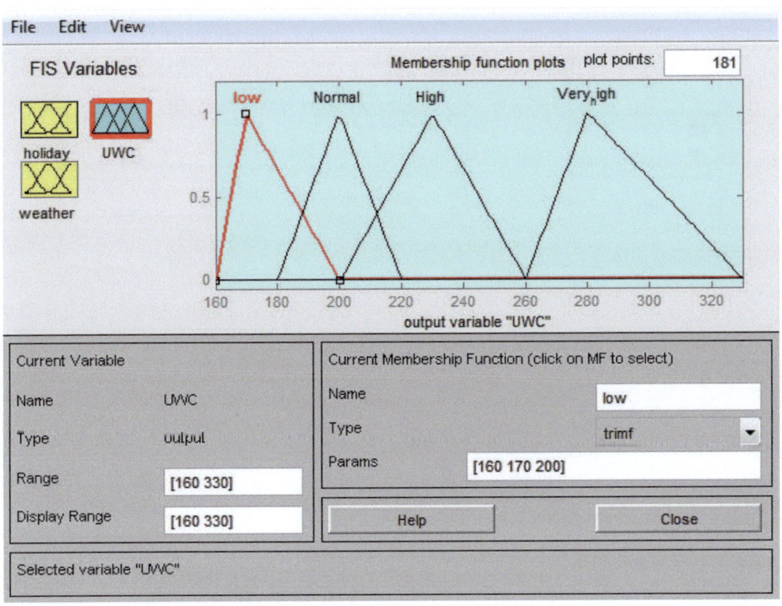

Fig. 7.12 Fuzzy output variables for Example 7.3

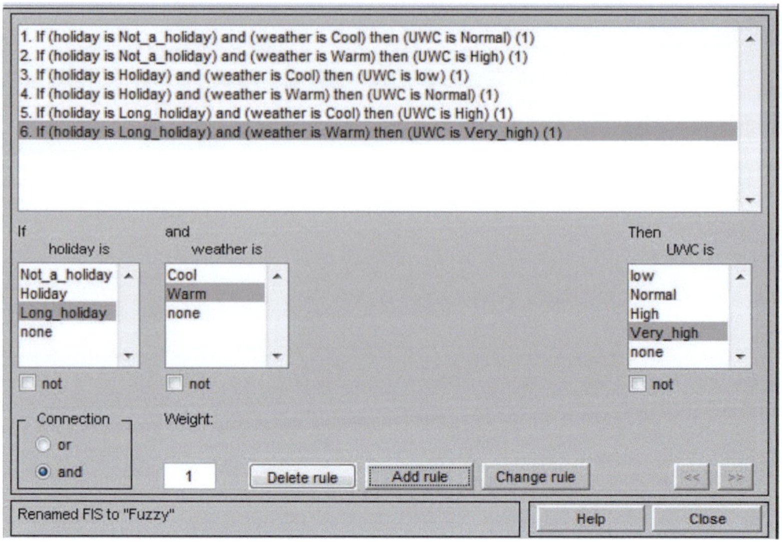

Fig. 7.13 Editing rules in the FIS

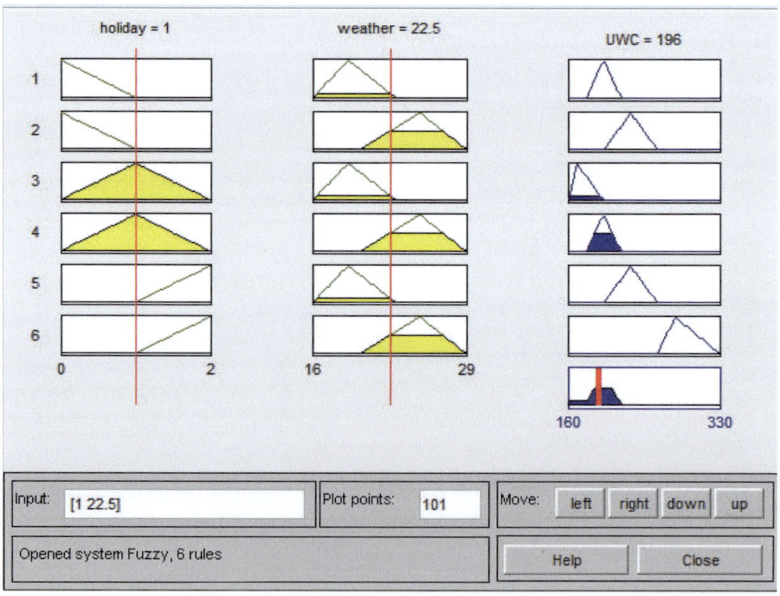

Fig. 7.14 Fuzzy inference system for Example 7.3

ANFIS uses a hybrid learning algorithm to identify parameters of *Sugeno*-type fuzzy inference systems. It applies a combination of the least-squares method and the backpropagation gradient descent method (as discussed in Chap. 6) for training FIS membership function parameters to emulate a given training data set.

A set of following commands are used to develop an ANFIS:
Input and output (target) data are introduced as

```
trnData = [X Y];
```

It is needed to select the number of membership functions (q), which is going to be considered through the fuzzy inference system. Furthermore, the type of the membership functions is selected in this step.

```
numMFs = q;
mfType = 'gbellmf';
```

Different types of membership functions as introduced in Table 7.2 could be used in the above syntax. Please note that in case of m-dimensional input and n-dimensional output, the number of rules in the fuzzy inference system of ANFIS would be $m \times n \times q$ where q membership functions are considered.

The syntax `in_fis` collects the necessary inputs needed to set a fuzzy inference system:

```
in_fis = genfis1(trnData,numMFs,mfType);
```

Finally, `out_fis` trains ANFIS by the predefined inputs as well as the training data set.

```
out_fis = anfis(trnData,in_fis);
```

Since an ANFIS is trained, different outputs could be simulated associated to the specific inputs of Z by the following syntax.

```
answer=evalfis(Z,out_fis)
```

Example 7.4: Fuzzy Inference System Modeling

Table 7.10 shows illustrative data of two-dimensional input and one-dimensional output. Develop an ANFIS to generalize the data. To validate the model, a separate set of data is presented in another table (Table 7.11).

Table 7.10 Calibration data set for Example 7.4

*X*1	*X*2	*Y*cal
24	110	10
26	75	20
32	105	30
35	65	40
45	25	50
50	20	60
100	40	70
54	20	80
123	16	90
30	70	45
42	28	55
48	18	66
105	45	70
58	22	78
120	18	89
20	100	9
30	80	45
32	105	30
60	25	82
110	18	86

Table 7.11 Validation data set for Example 7.4

*Z*1	*Z*2	*Y* value
33	80	48
34	106	32
31	62	43
41	27	47
44	22	56
102	30	75
55	25	72

Solution

The following commands are used to solve the problem:

```
trnData = [X Y];
numMFs = 2;
mfType = 'gbellmf';
in_fis = genfis1(trnData,numMFs,mfType);
out_fis = anfis(trnData,in_fis);
answer=evalfis(X,out_fis)
```

It should be notified that *X* and *Y* are defined as input and dependent variables which should be defined for both calibration and validation data, separately. Tables 7.12, 7.13, and 7.14 demonstrate the results of simulation.

Table 7.12 Results for calibration data set

Y	Estimated Y
10	12.20
20	28.10
30	29.94
40	45.32
50	55.23
60	67.72
70	70.34
80	74.74
90	89.69
45	36.81
55	47.44
66	64.98
70	69.79
78	79.49
89	88.95
9	6.41
45	41.34
30	29.94
82	80.15
86	86.44

Table 7.13 Results for validation data set

Y value	Estimated Y value
48	48.69
32	30.39
43	31.06
47	45.66
56	54.61
75	79.42
72	73.63

Table 7.14 Summary results for calibration and validation data sets

Data set	Linear correlation between actual and simulated Y	Volume error (%)
Calibration	0.986	5.978
Validation	0.976	6.818

7.6 Fuzzy Regression

Fuzzy regression is a type of regression model that uses possibility theory instead of probability theory to estimate a dependent variable by one or more independent variables. In a fuzzy regression, the dependent variable is presented by a fuzzy number. It enables providing the possible range of estimated variable instead of

Fig. 7.15 The typical fuzzy parameter of a fuzzy linear regression

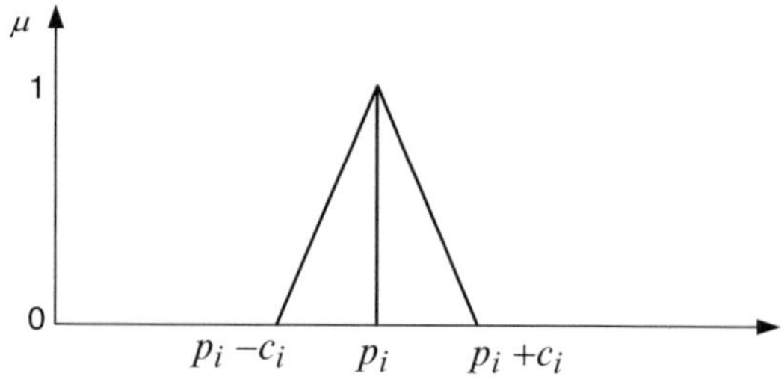

solely one value. It is especially suitable for cases that vagueness should be embedded in the final results of the model.

The general form of a multiple linear regression is

$$y = a_0 + a_1 x_1 + \cdots + a_n x_n \tag{7.8}$$

The general form of a fuzzy linear regression is written as

$$\widetilde{Y} = \widetilde{A}_0 + \widetilde{A}_1 x_1 + \widetilde{A}_2 x_2 + \cdots + \widetilde{A}_n x_n \tag{7.9}$$

where the parameters and the output of the model are both fuzzy numbers instead of the classic ones. The fuzzy parameters (coefficients) of the model could be represented by a triangular membership function with the center of p_i and the spread of c_i as shown in Fig. 7.15.

In this case, the membership of each number, a_i in range of $p_i - c_i \leq a_i \leq p_i + c_i$, is obtained as

$$\mu_{\widetilde{A}_i}(a_i) = \begin{cases} 1 - \dfrac{|p_i - a_i|}{c_i}; & p_i - c_i \leq a_i \leq p_i + c_i \\ 0; & \text{otherwise} \end{cases} \tag{7.10}$$

The results of multiplying the fuzzy coefficients by the input variables give a fuzzy number as output:

$$\widetilde{Y} = (p_0, c_0) + (p_1, c_1) x_1 + (p_2, c_2) x_2 + \cdots + (p_n, c_n) x_n \tag{7.11}$$

Figure 7.16 shows the fuzzy number of the output where the membership of each output value is calculated by Eq. 7.12.

$$\mu_{\widetilde{Y}}(y) = \begin{cases} 1 - \dfrac{\left| y - p_0 - \sum_{i=1}^{n} p_i x_i \right|}{c_0 + \sum_{i=1}^{n} c_i |x_i|}; & x_i \neq 0 \\ 1; & x_i = 0, \quad y = 0 \\ 0; & x_i = 0, \quad y \neq 0 \end{cases} \tag{7.12}$$

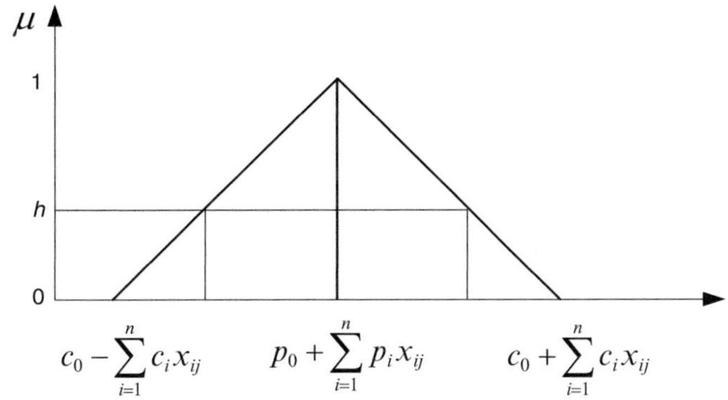

Fig. 7.16 The typical fuzzy output of a fuzzy linear regression

One of the most popular fuzzy regression methods uses the assumption that the membership of each output variable should be greater than the specific value of h or

$$\mu_{\tilde{Y}_j}\left(y_j\right) \geq h \tag{7.13}$$

as shown in Fig. 7.16.

In fact we try to find the regression parameters in a way that the output is limited to the following ranges:

$$p_0 + \sum_{i=1}^{n} p_i x_{ij} - (1 - h)\left[c_0 + \sum_{i=1}^{n} c_i x_{ij}\right] \leq y_j \tag{7.14}$$

$$p_0 + \sum_{i=1}^{n} p_i x_{ij} + (1 - h)\left[c_0 + \sum_{i=1}^{n} c_i x_{ij}\right] \geq y_j \tag{7.15}$$

These equations could be the constraints of an optimization problem where the objective function is to minimize the spread of the outputs. The objective function is presented as

$$\text{minimize}: \quad mc_0 + \sum_{j=1}^{m} \sum_{i=1}^{n} c_i \left|x_{ij}\right| \tag{7.16}$$

The procedure of fuzzy regression modeling using the optimization problem with the objective function of Eq. 7.16 and constraints of Eqs. 7.14 and 7.15 is demonstrated in the following example.

Example 7.5: Fuzzy Regression Model for Discharge Estimation

Table 7.15 Data presented for Example 7.5

Year	1	2	3	4	5	6
Rainfall	60	80	100	120	140	160
Discharge	100	120	130	160	230	240

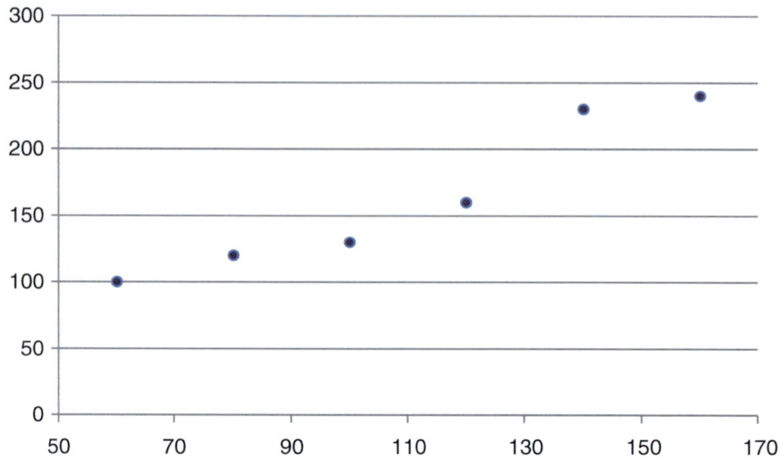

Fig. 7.17 Plot of rainfall (the x axis) versus discharge (the y axis) in Example 7.5

Use a fuzzy regression to model annual discharge of a basin to its annual rainfall as recorded in Table 7.15.

Solution

The following plot shows the input variable versus the dependent variable (Fig. 7.17).

As demonstrated by the data, the fuzzy regression contains one variable of six records, so the parameters of the model are

n = number of input variables = 1
m = number of observations = 6
unknown variables = c_0 and p_0 as well as c_1 and p_1

The objective function of fuzzy regression is

$$\text{minimize}: \quad 6c_0 + \sum_{j=1}^{6}\sum_{i=1}^{1} c_i |x_{ij}| \tag{7.17}$$

or in extended form

$$\text{minimize}: \quad 6c_0 + c_1x_{11} + c_1x_{12} + c_1x_{13} + c_1x_{14} + c_1x_{15} + c_1x_{16} \tag{7.18}$$

Considering $h = 0.7$, the constraints are

$$p_0 + \sum_{i=1}^{1} p_i x_{ij} - (1 - 0.7)\left[c_0 + \sum_{i=1}^{1} c_i x_{ij}\right] \le y_j \tag{7.19}$$

and

$$p_0 + \sum_{i=1}^{1} p_i x_{ij} + (1 - 0.7)\left[c_0 + \sum_{i=1}^{1} c_i x_{ij}\right] \ge y_j \tag{7.20}$$

which tends to

$$\begin{aligned}
p_0 + p_1 x_{11} - 0.3c_0 - 0.3c_1 x_{11} \le y_1 \\
p_0 + p_1 x_{12} - 0.3c_0 - 0.3c_1 x_{12} \le y_2 \\
p_0 + p_1 x_{13} - 0.3c_0 - 0.3c_1 x_{13} \le y_3 \\
p_0 + p_1 x_{14} - 0.3c_0 - 0.3c_1 x_{14} \le y_4 \\
p_0 + p_1 x_{15} - 0.3c_0 - 0.3c_1 x_{15} \le y_5 \\
p_0 + p_1 x_{16} - 0.3c_0 - 0.3c_1 x_{16} \le y_6
\end{aligned} \tag{7.21}$$

and

$$\begin{aligned}
p_0 + p_1 x_{11} + 0.3c_0 + 0.3c_1 x_{11} \ge y_1 \\
p_0 + p_1 x_{12} + 0.3c_0 + 0.3c_1 x_{12} \ge y_2 \\
p_0 + p_1 x_{13} + 0.3c_0 + 0.3c_1 x_{13} \ge y_3 \\
p_0 + p_1 x_{14} + 0.3c_0 + 0.3c_1 x_{14} \ge y_4 \\
p_0 + p_1 x_{15} + 0.3c_0 + 0.3c_1 x_{15} \ge y_5 \\
p_0 + p_1 x_{16} + 0.3c_0 + 0.3c_1 x_{16} \ge y_6
\end{aligned} \tag{7.22}$$

The following command solves the above linear optimization problem, where f = objective function, A = matrix of coefficients of inequity equations, and B = matrix of the right-hand limit of the inequity equations. x contains the unknown variables of the objective function:

```
x=linprog(f,A,B);
```

f is a $[2 \times (n + 1), 1]$ matrix containing the coefficient of the objective function. The first half of the elements of the matrix contains 0 as p parameter does not appear in the objective function. The next element is equal to m, and the n remainder values are in fact $\sum_{j}=1^m x_{ij}$. For the present exercise, f would be inequity equations

$$\begin{bmatrix} 0 \\ 0 \\ 6 \\ 660 \end{bmatrix}$$

A is a $[2 \times m, (n + 3)]$ matrix containing the coefficients of the constraints. The first half of the elements of the matrix presents the coefficients of the equations set of 7.21

and the second half presents the coefficients of the equation set of 7.22. The first column of the matrix contains 1 in the first half and −1 in the second half, the second column contains x_{ij} values in its first half and $-\ x_{ij}$ in its second half, the third column contains $-\ (1\ -\ h)$, and the first half of the fourth column later on contains $-\ (1\ -\ h)x_{ij}$ duplicating in the second half of the column. For the present example, the A matrix is

$$A = \begin{bmatrix} +1 & 60 & -0.3 & -18 \\ +1 & 80 & -0.3 & -24 \\ +1 & 100 & -0.3 & -30 \\ +1 & 120 & -0.3 & -36 \\ +1 & 140 & -0.3 & -42 \\ +1 & 160 & -0.3 & -48 \\ -1 & -60 & -0.3 & -18 \\ -1 & -80 & -0.3 & -24 \\ -1 & -100 & -0.3 & -30 \\ -1 & -120 & -0.3 & -36 \\ -1 & -140 & -0.3 & -42 \\ -1 & -160 & -0.3 & -48 \end{bmatrix}$$

B is the right-hand limit of the inequity equations. It is a $[2 \times m,\ 1]$ matrix containing y_j values for the first half and $-\ y_j$ for the second half of the matrix. For the present exercise, the B matrix would be

$$B = \begin{bmatrix} 100 \\ 120 \\ 130 \\ 160 \\ 230 \\ 240 \\ -100 \\ -120 \\ -130 \\ -160 \\ -230 \\ -240 \end{bmatrix}$$

Executing the `x=linprog(f,A,B)` command will result in the following answers:

```
x =
```

```
-8.7500
 1.5625
37.5000
 0.2083
```

which means

$$(p_0, c_0) = (-8.75, 37.5)$$

and

$$(p_1, c_1) = (1.56, 0.2)$$

Changing h to 0.3, the A matrix and results will be

$$A = \begin{bmatrix} +1 & 60 & -0.7 & -42 \\ +1 & 80 & -0.7 & -56 \\ +1 & 100 & -0.7 & -70 \\ +1 & 120 & -0.7 & -84 \\ +1 & 140 & -0.7 & -98 \\ +1 & 160 & -0.7 & -112 \\ -1 & -60 & -0.7 & -42 \\ -1 & -80 & -0.7 & -56 \\ -1 & -100 & -0.7 & -70 \\ -1 & -120 & -0.7 & -84 \\ -1 & -140 & -0.7 & -98 \\ -1 & -160 & -0.7 & -112 \end{bmatrix}$$

and

```
x =
```

```
-8.7500
 1.5625
16.0714
 0.0893
```

which means

$$(p_0, c_0) = (-8.75, 16.07)$$

and

$$(p_1, c_1) = (1.56, 0.09)$$

For $h = 0.1$, the A matrix and results will be

$$A = \begin{bmatrix} +1 & 60 & -0.9 & -54 \\ +1 & 80 & -0.9 & -72 \\ +1 & 100 & -0.9 & -90 \\ +1 & 120 & -0.9 & -108 \\ +1 & 140 & -0.9 & -126 \\ +1 & 160 & -0.9 & -144 \\ -1 & -60 & -0.9 & -54 \\ -1 & -80 & -0.9 & -72 \\ -1 & -100 & -0.9 & -90 \\ -1 & -120 & -0.9 & -108 \\ -1 & -140 & -0.9 & -126 \\ -1 & -160 & -0.9 & -144 \end{bmatrix}$$

```
x =
```

```
 -8.7500
  1.5625
 12.5000
  0.0694
```

and finally for $h = 0.0$, the results are

```
x =
```

```
 -8.7500
  1.5625
 11.2500
  0.0625
```

Table 7.16 shows different results for Example 7.5 obtained by different values of h.

Table 7.16 Different results for Example 7.5 according to different values of h

Actual dependent variable, Y	Estimated Y	Maximum estimation of Y	Minimum estimation of Y
$h = 0.7$			
100	85	134.998	35.002
120	116.25	170.414	62.086
130	147.5	205.83	89.17
160	178.75	241.246	116.254
230	210	276.662	143.338
240	241.25	312.078	170.422
$h = 0.3$			
100	85	106.429	63.5706
120	116.25	139.465	93.0346
130	147.5	172.501	122.496
160	178.75	205.537	151.962
230	210	238.573	181.426
240	241.25	271.609	210.890
$h = 0.1$			
100	85	101.664	68.336
120	116.25	134.302	98.198
130	147.5	166.94	128.06
160	178.75	199.578	157.922
230	210	232.216	187.784
240	241.25	264.854	217.646
$h = 0.0$			
100	85	100	70
120	116.25	132.5	100
130	147.5	165	130
160	178.75	197.5	160
230	210	230	190
240	241.25	262.5	220

The maximum and minimum estimations obtained by $h = 0.0$ are actually obtained by lower and upper bounds of data as shown in Fig. 7.18.

7.7 Summary

After presenting basic information on the fuzzy logic, the chapter discussed three major applications of fuzzy theory in the field of water resources and environmental engineering. Fuzzy clustering, fuzzy inference system, and fuzzy regression are three subjects discussed in the chapter.

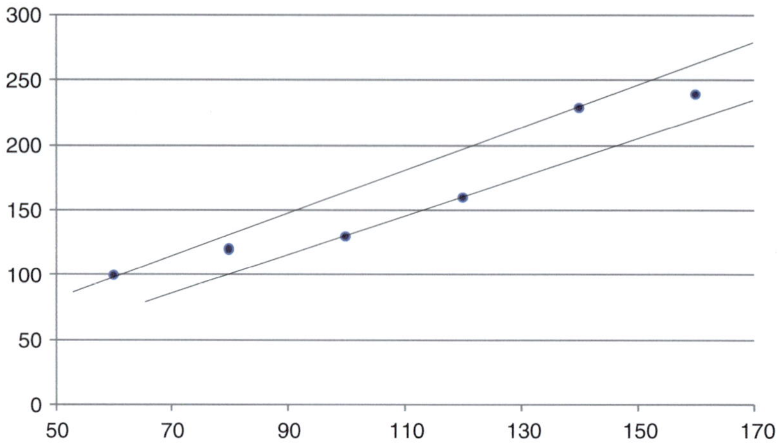

Fig. 7.18 Lower and upper regression estimation of Example 7.5 according to fuzzy regression by $h = 0.0$

Fuzzy clustering allows each feature vector to belong to more than one cluster with different membership degrees (between 0 and 1) and vague or fuzzy boundaries between clusters. Fuzzy C-means (FCM) is a data clustering technique wherein each data point belongs to a cluster to some degree that is specified by a membership grade.

Fuzzy inference system (FIS) is usually used for mapping a set of input/output data in which there is no predetermined structure. FIS is presented as logical rules named as *if–then* rules. The possibility logic of fuzzy systems is a useful tool for considering the vagueness in modeling real-world problems. What makes the use of FIS more efficient than regression-based methods in some cases is to deal with the vagueness in predictors and the nonlinear relationship between the predictors and dependent variables. Furthermore, FIS can deal with both linguistic and quantitative variables in the process of modeling. In contrast to the conventional data-driven methods, which try to find a logical relationship between input and output variables from the observed data, FIS gets benefit from both the concept of the problem and the information within the observed data. Two types of *Mamdani* and *Sugeno* fuzzy inference systems were presented in the chapter. Adaptive neuro-fuzzy inference system (ANFIS) is a kind of neural network that is based on *Sugeno* fuzzy inference system. Since it integrates both neural networks and fuzzy logic principles, it has potential to capture the benefits of both in a single framework. Its inference system corresponds to a set of fuzzy *if–then rules* that have learning capability to approximate nonlinear functions. ANFIS uses a hybrid learning algorithm to identify parameters of *Sugeno*-type fuzzy inference systems. It applies a combination of the least-squares method and the backpropagation gradient descent method for training FIS membership function parameters to emulate a given training data set.

Fuzzy regression is a type of regression model that uses possibility theory instead of probability theory to estimate a dependent variable by one or more independent variables. In a fuzzy regression the dependent variable is presented by a fuzzy number. It enables providing the possible range of estimated variable instead of solely one value. It is especially suitable for cases that vagueness should be embedded in the final results of the model.

Workshop

The following data shows the results of an optimization model for reservoir operation, which shows optimum release of a reservoir during a drought in a specific season. The inputs of the model include inflow to the reservoir as well as the storage volume at the beginning of the month (in million cubic meters). Derive an ANFIS model to generalize the operating rules for the reservoir (Tables 7.17 and 7.18).

Solution

Before modeling the input/output data, a brief discussion on the concept of the operation rule of this example is presented. The concept of optimum release of the reservoir is in fact based on the well-known hedging rule for reservoir operation during droughts. A two-point hedging rule defining the rules of this example is demonstrated in Fig. 7.19. Sum of inflow to the reservoir and the storage volume is considered as the drought index which activates three different rules due to the Trigger 1 and Trigger 2.

Table 7.17 Calibration data set presented for the workshop

Inflow to the reservoir	Storage volume at the beginning of the month	Optimum release
50	50	100.0
176	90	161.5
150	60	143.0
130	100	149.6
190	120	170.0
230	100	170.0
200	120	170.0
80	80	126.5
50	40	90.0
60	89	122.9

Table 7.18 Validation data set presented for the workshop

Inflow	Storage volume	Optimum release
70	100	129.8
90	120	143
230	130	170
60	45	105

Fig. 7.19 A schematic hedging rule for reservoir operation during drought

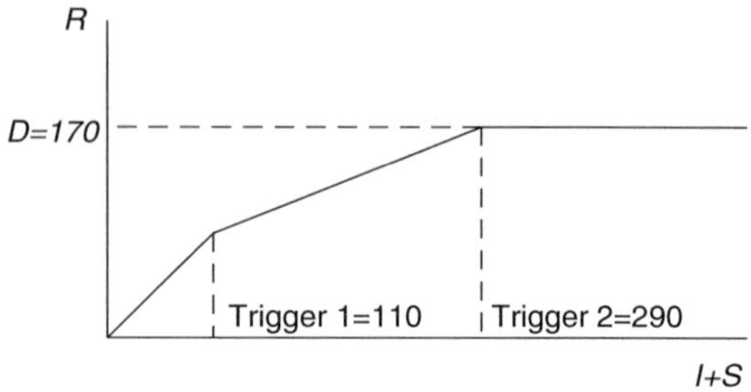

The rules associated with the drought index (*I* + *S*) and drought triggers are presented as follows:

$$\begin{array}{lll} R = I + S & \text{if} & I + S \leq \text{Trigger } 1 = 110 \\ R = 0.33(I + S) + 73.7 & \text{if} & \text{Trigger } 1 \leq I + S \leq \text{Trigger } 2 \\ R = D = 170 & \text{if} & I + S \geq \text{Trigger } 2 = 290 \end{array}$$

where *R* is the release from the reservoir in the specific season and *D* is the water demand at the same time. The major difference between the hedging rule (HR) and the conventional rules of reservoir operation is that the HR avoids full allocation of water to the demands even in case that the available water is equal or more than the demand. HR tries to save water to prevent the severe drought in near future. In fact HR prefers less sever but longer droughts instead of short droughts with a high severity.

The following commands show the short program developed to map 10 by 1 matrix of output, *Y*, by 10 by 2 matrix of inputs, *X*. Furthermore, 4 by 2 matrix of *Xval* is considered for the model's validation:

```
trnData = [X Y];
numMFs = 2;
mfType = 'gbellmf';
in_fis = genfis1(trnData,numMFs,mfType);
out_fis = anfis(trnData,in_fis);
answer=evalfis(Xval,out_fis)
```

It should be notified that number of membership functions and their type could be changed alternatively.

The results of calibration and validation are tabulated below. As demonstrated in Table 7.19, ANFIS has performed as an exact estimator in calibration and has produced minor errors in validation.

Table 7.19 Results obtained by ANFIS in calibration and validation

Actual release	Estimated release
100.0	100.0
161.5	161.5
143.0	143.0
149.6	149.6
170.0	170.0
170.0	170.0
170.0	170.0
126.5	126.5
90.0	90.0
122.9	122.9
129.8	130.9
143	151.7
170	172.2
105	98.0

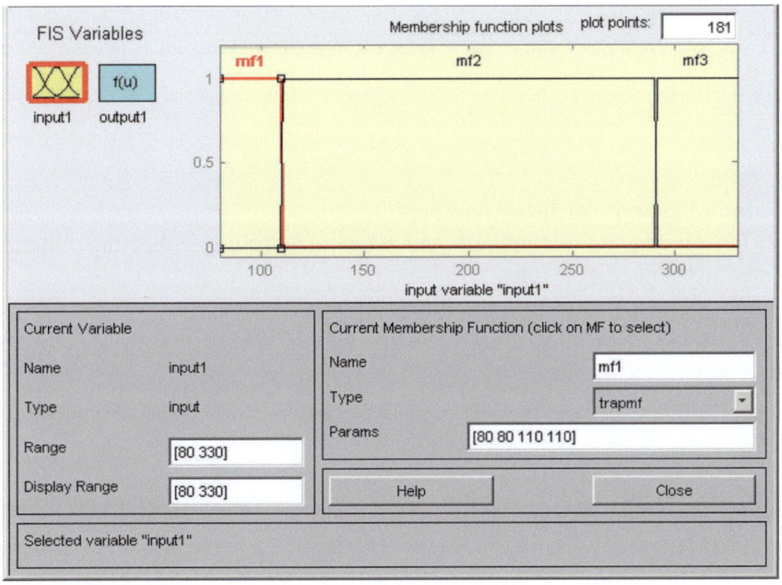

Fig. 7.20 Defining parameters and type of input membership functions

Now we would like to set up a fuzzy inference system based on the concept of the reservoir operating rule as discussed by the hedging rule. As the output of each rule is presented by a linear regression, the *Sugeno*-type FIS is an appropriate system for setting up the model. Figures 7.20, 7.21, and 7.22 demonstrate different steps of developing this system.

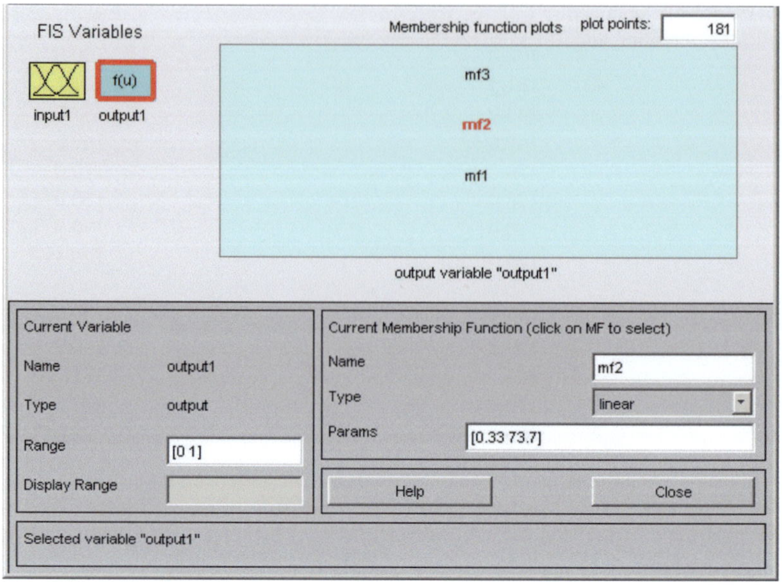

Fig. 7.21 Defining parameters and type of output membership functions

Step 1: Introducing input fuzzy numbers

Step 2: Introducing the parameters and types of output membership functions as defined by the hedging rules

Step 3: Developing and testing the fuzzy inference system

a

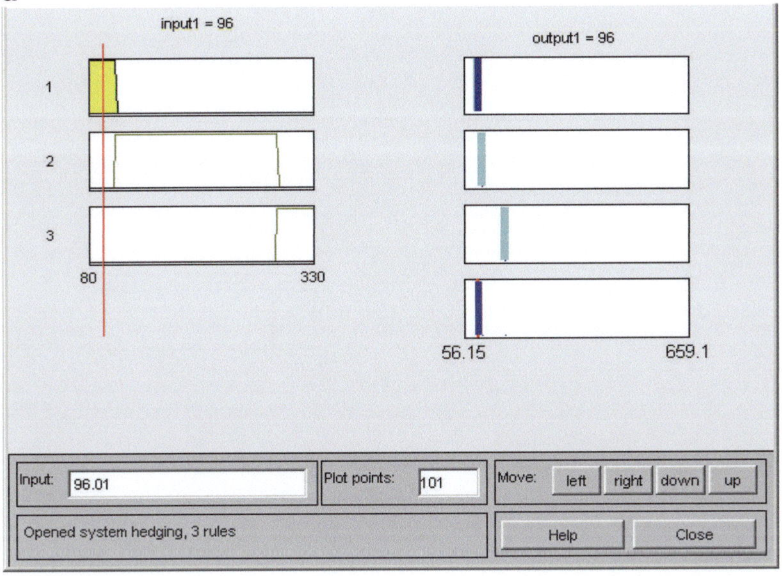

b

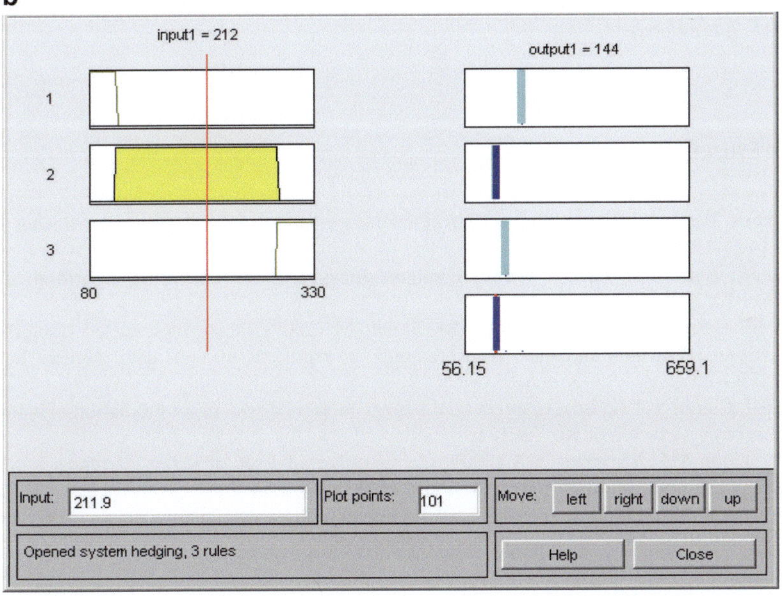

Fig. 7.22 *Sugeno*-based fuzzy inference system for the workshop with examples of inputs less than Trigger 1 (**a**), between Trigger 2 and Trigger 1 (**b**), and more than Trigger 2 (**c**)

c

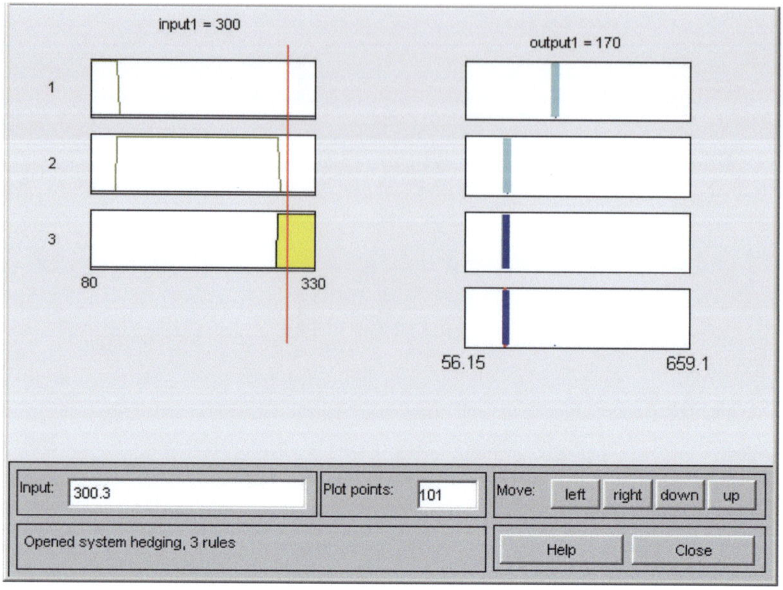

Fig. 7.22 (continued)

References

Bardossy A, Bogardi I, Duckstein L (1990) Fuzzy regression in hydrology. Water Resour Res 26 (7):1497–1508

Bezdek JC, Ehrlich R, Full W (1984) Fcm: the fuzzy c-means clustering algorithm. Comput Geosci 10(2–3):191–203

Casper M, Gemmar P, Gronz O, Johst M, Stuber M (2007) Fuzzy logic-based rainfall-runoff modelling using soil moisture measurements to represent system state. Hydrol Sci J 52 (3):478–490

Chang F-J, Chang Y-T (2006) Adaptive neuro-fuzzy inference system for prediction of water level in reservoir. Adv Water Res 29:1–10

Firat M, Turan ME, Yurdusev MA (2009) Comparative analysis of fuzzy inference systems for water consumption time series prediction. J Hydrol 374:235–241

Kucukmehmetoglu M, Şen Z, Özger M (2010) Coalition possibility of riparian countries via game theory and fuzzy logic models. Water Resour Res 46:W12528

Lee M, McBean EA, Ghazali M, Schuster CJ, Huang JJ (2009) Fuzzy-logic modeling of risk assessment for a small drinking-water supply system. J Water Resour Plann Manage 135 (6):547–552

Lohani AK, Goel NK, Bhatia KKS (2007) Deriving stage–discharge–sediment concentration relationships using fuzzy logic. Hydrol Sci J 52(4):793–807

Mathon BR, Ozbek MM, Pinder GF (2008) Transmissivity and storage coefficient estimation by coupling the Cooper–Jacob method and modified fuzzy least-squares regression. J Hydrol 353:267–274

Moghaddamnia A, Ghafari Gousheh M, Piri J, Amin S, Han D (2009) Evaporation estimation using artificial neural networks and adaptive neuro-fuzzy inference system techniques. Adv Water Res 32:88–97

Özelkan EC, Ni F, Duckstein L (1996) Relationship between monthly atmospheric circulation patterns and precipitation: fuzzy logic and regression approaches. Water Resour Res 32 (7):2097–2103

Ren L, Xiang X-Y, Ni J-J (2013) Forecast modeling of monthly runoff with adaptive neural fuzzy inference system and wavelet analysis. J Hydrol Eng 18(9):1133–1139

Shrestha RR, Simonovic SP (2010) Fuzzy nonlinear regression approach to stage-discharge analyses: case study. J Hydrol Eng 15(1):49–56

Shu C, Ouarda TBMJ (2008) Regional flood frequency analysis at ungauged sites using the adaptive neuro-fuzzy inference system. J Hydrol 349:31–43

Talei A, Chua LHC, Wong TSW (2010) Evaluation of rainfall and discharge inputs used by Adaptive Network-based Fuzzy Inference Systems (ANFIS) in rainfall–runoff modeling. J Hydrol 391:248–262

Terzi Ö, Keskin ME, Taylan ED (2006) Estimating evaporation using ANFIS. J Irrig Drain Eng 132(5):503–507

Wang K-H, Altunkaynak A (2012) A comparative case study of rainfall-runoff modeling between SWMM and FUZZY logic approach. J Hydrol Eng 17(2):283–291

Chapter 8
Hybrid Models and Multi-model Data Fusion

Abstract The need for increased accuracy and precision in data-driven models has motivated the researchers to develop innovative models. Hybrid models and multi-model ensemble estimations are applied to increase accuracy and precision of single models. To get an idea about how different models could be combined in a way to increase each other's abilities, the chapter begins with a summary on the characteristics of the models presented in the previous chapters of the book. The models are compared based on different criteria to give the readers ideas on how to take advantages of the models' strengths and avoid their weakness through the hybrid models and multi-model data fusion approach. The chapter continues with the examples of hybrid models and general techniques of multi-model data fusion. The approach of multi-model data fusion contains an important process of individual model generation which is going to be discussed in the last section of the chapter.

Keywords Hybrid models • Multi-model data fusion • Stacking • Individual model generation

8.1 Introduction

The need for increased accuracy and precision in data-driven models has motivated the researchers to develop innovative models. Actually, hybrid models and multi-model ensemble estimations are applied to meet these specifications and to integrate strengths of single models. In other words, the main aim of developing hybrid models is to integrate the advantages of two or more models in a way to improve the capability of single ones. The hybrid model is usually a combination of models in series. Conventionally, one model among the others is considered as the main model, and the others play the role of preprocessing or postprocessing techniques. Data fusion is an emerging area of research that covers a broad range of application areas. The principal objective of data fusion is to provide a solution that either is

S. Araghinejad, *Data-Driven Modeling: Using MATLAB® in Water Resources and Environmental Engineering*, Water Science and Technology Library 67,
DOI 10.1007/978-94-007-7506-0_8,

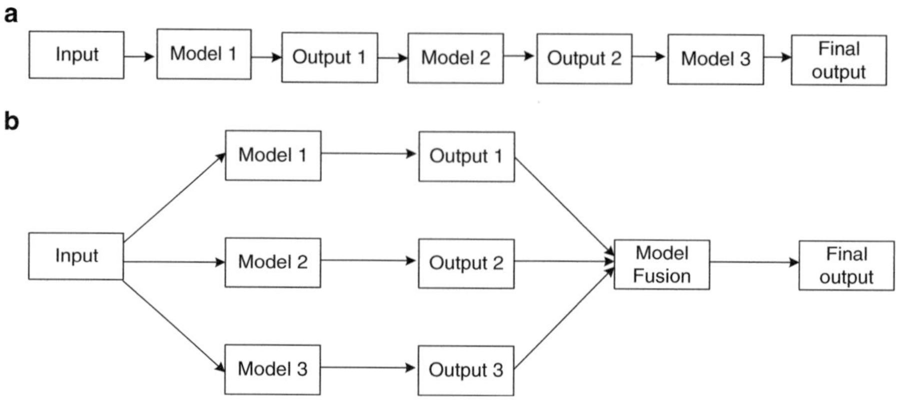

Fig. 8.1 Schematic algorithms of hybrid models (**a**) and multi-model data fusion (**b**) approaches

more accurate or offers additional inferences more than those that could be obtained through the use of individual data. Data fusion provides new modeling opportunities in the water resource and environmental fields. Model fusion might benefit from the ability to combine information derived from multiple sources, such as the individual outputs from different simulation models. Data fusion researches are divided into two broad groups. The first takes the view that data fusion is the amalgamation of raw information to produce an output, while the second makes use of a more generalized view of data fusion in which both raw and processed information can be fused into useful outputs including higher-level decision.

Recently, researchers such as See and Abrahart (2001), Abrahart and See (2002), and Shu and Burn (2004) have used model fusion approaches in hydrological engineering. See and Abrahart (2001) used data fusion approach for continuous river-level forecasting where data fusion was the amalgamation of information from multiple sensors and different data sources. Abrahart and See (2002) evaluated six data fusion strategies and found that data fusion by an artificial neural network (ANN) model provided the best solution. Shu and Burn (2004) applied artificial neural network ensembles in pooled flood frequency analysis for estimating the index flood and the 10-year flood quintiles. The data fusion method was used to combine individual ANN models in order to enhance the final estimation.

Figure 8.1 demonstrates the general structure of hybrid models as well as the model fusion approach. As shown in the figure, different individual models in the process of hybrid models contribute in different levels of the modeling process, while, in a multi-model data fusion approach, all individual models might have a chance to contribute in the final output.

The contents of this chapter are structured as presented in Fig. 8.2. As it is demonstrated in the figure, the chapter begins with a summary on the characteristics of the models presented in this book. The models are compared in that section based on different criteria to give the readers ideas on how to take advantages of the models' strengths and avoid their weakness through the mentioned models. The chapter continues with the examples of hybrid models and general techniques

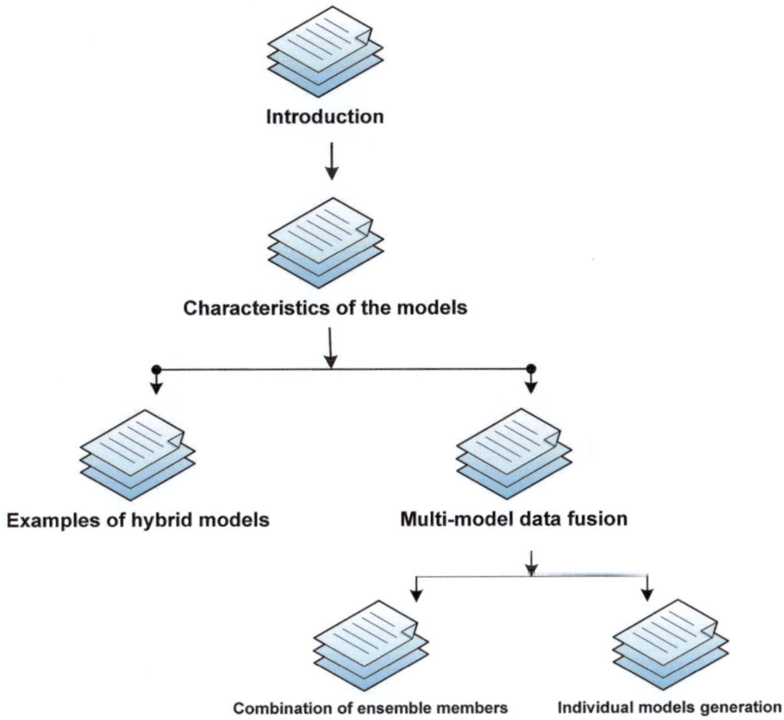

Fig. 8.2 Structure of this chapter

of data fusion. The approach of multi-model data fusion contains an important process of individual model generation which is going to be discussed in the last section of the chapter.

8.2 Characteristics of the Models

There is no general stepwise algorithm for developing a hybrid model. The basic ideas of hybrid models are usually originated by the experience and knowledge of the modeler. Of course, the review of the well-known existing hybrid models could also be very helpful for a modeler to develop his/her own hybrid model. To understand better the idea of these models and getting ideas on how the combined models could improve the capability of single ones, it may be useful to review the characteristics of the most famous data-driven models.

To get an idea about how different models could be combined in a way to increase each other's abilities, a summary of the applicable fields where different models could be applied is shown in Table 8.1. The applicable fields are considered as event modeling (like the application of linear regression and MLP network as

Table 8.1 Different fields of applications for data-driven models

Models	Event modeling	Time series modeling	Spatial analysis	Clustering/classification
ANFIS	√			√
ARIMA		√		
ARMA		√		
ARMAX		√		
FCM				√
FIS	√			√
FR	√		√	
GRNN	√	√	√	
IDNN	√	√		
K-NN	√	√	√	√
MAR		√		
MLP	√		√	√
MLR	√		√	
NLR	√		√	
PNN				√
RBE	√		√	
RNN	√			
SVM	√			√
TDRNN	√	√		

discussed in Chaps. 3 and 5), time series modeling (as discussed in Chap. 4), spatial analysis, and the field of clustering/classification. Furthermore, the selected strength and weakness of models are reviewed in Table 8.2.

For simplification, the following expression reviews the abbreviation of the models, which have been reviewed in the previous:

ANFIS Adaptive neuro-fuzzy inference system
ARIMA Autoregressive integrated moving average
ARMA Autoregressive moving average model
ARMAX Autoregressive moving average with eXogenous input
FCM Fuzzy C-means
FIS Fuzzy inference system
FR Fuzzy regression
GRNN Generalized regression neural network
IDNN Input delay neural network
K-NN K-nearest neighbor regression
MAR Multivariate autoregressive model
MLP Multilayer perceptron
MLR Multiple linear regression model
NLR Nonlinear regression model
PNN Probabilistic neural network
RBE Radial basis estimator
RNN Recurrent neural network
SVM Support vector machine
TDRNN Time-delay neural network

Table 8.2 Strengths and weaknesses of different models

Models	Strengths	Weaknesses
ANFIS	Benefits from the physical concept of a system by if-then rules	Needs adequate data to represent different patterns and rules observed through the life of a system
ARIMA	Can be used for a time series with a slight trend	Cannot be used for time series data generation
ARMA	Can model the temporal correlation between data	Cannot be used for a multivariable problem
ARMAX	Considers an extra variable except the information within a time series	Can use only one variable as external variable of modeling
FCM	Is one of the best clustering methods	Is only applicable to the field of clustering
FIS	Can use descriptive data in the modeling process	Does not use a systematic way for calibration
FR	Provides outputs by the possibility logic (fuzzy numbers), explicitly	Is not suitable for nonlinear problems
GRNN	Offers a straightforward nonparametric regression	Not much flexible for function approximation as it contains only one parameter to be calibrated
IDNN	Can model time series by neural network methods	Needs more complicated calibration
K-NN	Could be used for both linear and nonlinear problems	Cannot be used for extrapolation
	Does not need any specific distribution to be applied in a case	Needs parameter calibration before being applied
MAR	Can model time series in a multivariate manner	Becomes too complex in case of higher order of the model
MLP	Acts flexible in fitting on the data	Does not have a straight forward calibration algorithm
MLR	Provides outputs in a probabilistic manner, explicitly	Increases estimation error significantly in case of increased nonlinearity in the structure of the problem
NLR	Suitable for nonlinear problems	Needs real data to have the form of a well-known mathematical function
PNN	One of the best classification models	Is applicable to the limited field of classification
RBE	Can be used as a spatial model	Is not much flexible for function approximation as it contains only one parameter to be calibrated
RNN	Considers recurrent structures in the process of modeling	Has calibration problems
SVM	More robust than neural networks in some cases	Some limitations in higher-order problems
TDRNN	Considers both temporal and recurrent structures in the process of modeling	Contains too many parameters to be calibrated comparing to the other networks

Table 8.3 Scores of different group of models based on the four different criteria

Models	Criterion	Score
Parametric regression models	Simplicity of calibration	• • •
	Flexibility	•
	Generalization	• •
	Areas of application	• •
Nonparametric regression models	Simplicity of calibration	• • • •
	Flexibility	• • •
	Generalization	• •
	Areas of application	• • • •
Autoregressive models	Simplicity of calibration	• • •
	Flexibility	•
	Generalization	• • • •
	Areas of application	•
Support vector machines	Simplicity of calibration	• • •
	Flexibility	• • •
	Generalization	• • • •
	Areas of application	• •
Artificial neural networks	Simplicity of calibration	•
	Flexibility	• • • •
	Generalization	• • •
	Areas of application	• • • •
Statistical ANNs	Simplicity of calibration	• • • •
	Flexibility	• • •
	Generalization	• •
	Areas of application	• • •
Fuzzy models	Simplicity of calibration	• •
	Flexibility	• • •
	Generalization	• • • •
	Areas of application	• • • •

In addition to the applicable fields of a model, they all have specific characteristics called their strengths and weaknesses. The ability or inability of modeling nonlinear systems and multivariate systems and uncertainty of processes and descriptive data are examples to determine the strengths and weaknesses of models. Furthermore, the simplicity and easiness of calibration of models and also their complexity of formulation are other important characteristics that increase or decrease the strengths of a model. The most well-known strengths and weaknesses of data-driven models are reviewed in Table 8.2.

The models presented in the previous chapters are categorized into seven families of *parametric regression* models, *nonparametric regression* models, *autoregressive* models, *artificial neural networks*, *statistical artificial neural networks*, *support vector machines*, and *fuzzy models*. In Table 8.3, these categories are compared by four criteria of "simplicity of calibration," "flexibility," "generalization," and "applicable areas." Simplicity of calibration is considered as a criterion for preference of a specific model over the others in cases that they perform more and less similar. *Flexibility* of a model refers to a criterion in which a model could be calibrated in a flexible manner. On the other hand, a flexible model is the one that can take different parameters in a way to be applied for different forms of observed data. For instance, a model could be calibrated to perform better for either extreme values of observed data or near-normal values of them. *Generalization* refers to ability where they can perform both in extrapolation and interpolation as well as in calibration and validation. Finally, the applicable field is considered as another criterion for a family of models.

8.3 Examples of Hybrid Models

This section reviews three examples of developing hybrid models. The examples explain how the characteristics of different models can combine to improve the ability of individual ones.

Example 8.1: Nonlinear Fuzzy Modeling

Develop a hybrid model to estimate an unknown variable by a nonlinear fuzzy model.

Solution
As described in Chap. 7, the presented fuzzy regression model can be used for *linear* fuzzy estimation. Meanwhile, an MLP model can be employed for nonlinear mapping function. So, we can combine these two models to estimate a variable in a nonlinear fuzzy manner. The proposed structure of this hybrid model is shown in Fig. 8.3. As it is demonstrated in the figure, the model has a feedforward architecture in which input data are mapped to the estimated data considering target value. The estimated data in combination with the real target data are considered as the inputs of a fuzzy regression model. The final system uses the original input data, X, and processes them to an estimated fuzzy output.

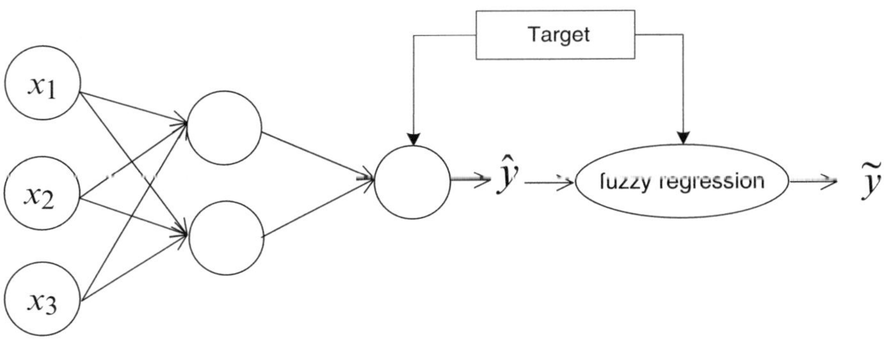

Fig. 8.3 A proposed hybrid model for nonlinear fuzzy estimation

Example 8.2: Mapping Different States of a System

In many cases of modeling in the field of water resources and environmental engineering, extreme value estimation is of most significance. The issue is that an individual model is rarely found to be much flexible for modeling maximum, minimum, and normal states. Develop a hybrid model to handle this issue.

Solution
Stepwise algorithm for developing such system is presented as follows:

Classify input/output data by m classes in which each class represents a specific state of data (say extreme values and near normal). m is 3 in Fig. 8.4.

Develop m different ANN models to map each class of data. Obviously, each individual ANN maps its associated set of data with less error than the other models.

Then, the working system of the proposed hybrid model is set up as shown in Fig. 8.4. As it is shown in the figure, each input data is classified by the PNN, first. The class of input data determines the ANN model which is applied to estimate the output associated to the input data.

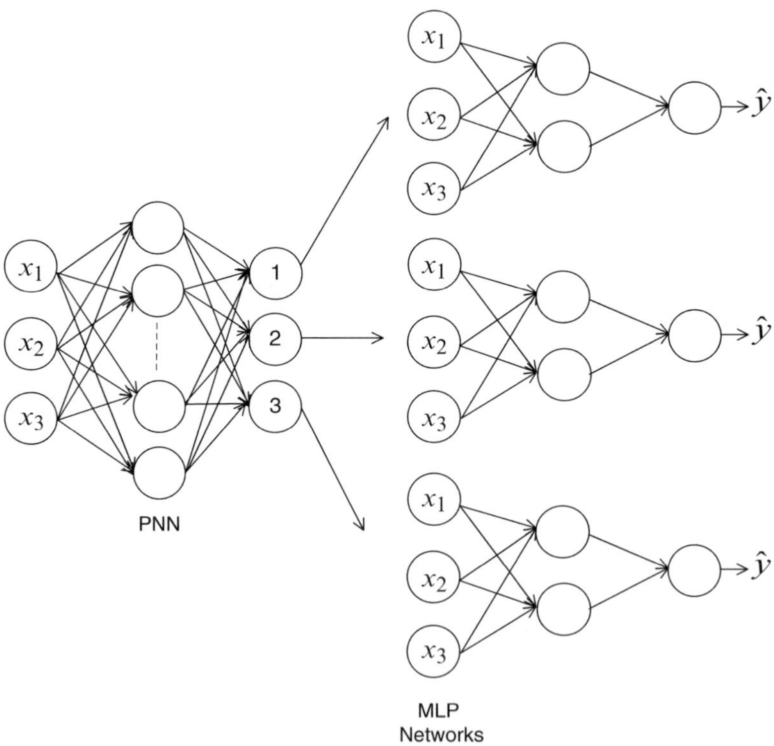

Fig. 8.4 A proposed hybrid model for mapping different states of a system

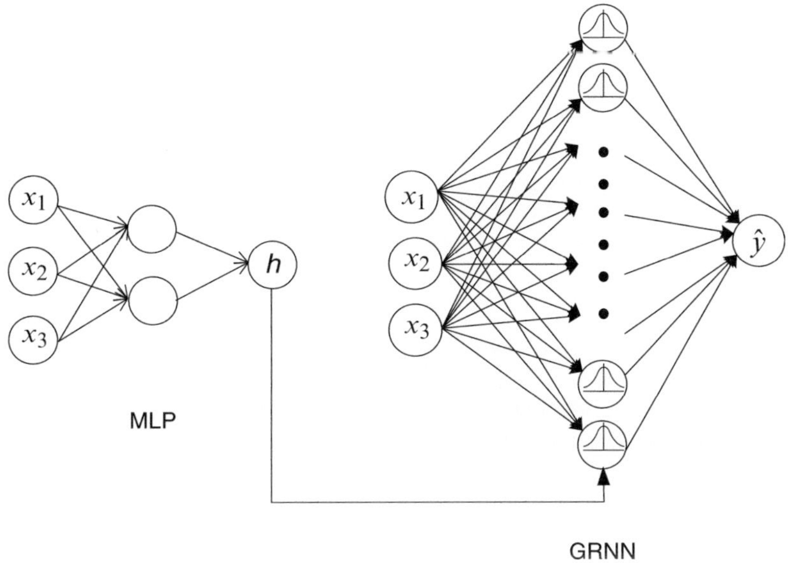

Fig. 8.5 A proposed hybrid model for applying a dynamic spread for the well-known GRNN model

Example 8.3: Dynamic Spread for Statistical Networks

Develop a statistical neural network (say GRNN) with a dynamic spread (h) according to changing different input variables. This will be a more flexible GRNN than the conventional form as it improves the estimation process by changing the spread parameter for different states of a problem.

Solution
The proposed hybrid model in this case is a combination of an MLP and a GRNN as shown in Fig. 8.6. For n observed data, n values of h are obtained in a way to better estimate each one the individual observations. The input observed data and the associated h are used to train an MLP. In real cases, the system is developed as shown in Fig. 8.5. The new input data are first introduced in the calibrated MLP to calculate the appropriate spread (h). The same input data as well as the selected h are introduced to the GRNN to estimate the final output.

8.4 Multi-model Data Fusion

A single data-driven model is represented by the following equation:

$$\hat{y}_i = f(X_i) + \varepsilon_i \quad i = 1, \ldots, n \tag{8.1}$$

where X = vector of predictors, $\hat{y}$ = dependent variable, ε = model error, and n = number of observed data. In the case of using multiple models to estimate y, considering similar inputs, Eq. 8.1 is changed to the following matrix form:

$$\left[\hat{Y}\right] = \begin{bmatrix} \hat{y}_{i1} \\ \hat{y}_{i2} \\ \vdots \\ \hat{y}_{im} \end{bmatrix} = \begin{bmatrix} f_1(X_i) \\ f_2(X_i) \\ \vdots \\ f_m(X_i) \end{bmatrix} + \begin{bmatrix} \varepsilon_{i1} \\ \varepsilon_{i2} \\ \vdots \\ \varepsilon_{im} \end{bmatrix} \quad i = 1, \ldots, n \tag{8.2}$$

where m = number of models used to estimate y and $\left[\hat{Y}\right]$ = matrix of estimations of y provided by different individual models. Using the data fusion approach, $\left[\hat{Y}\right]$ is summed up to a unique estimation of $\hat{y}$. The following paragraphs describe some of the general methods of data fusion. The model fusion techniques are divided into two steps. The first step is to create individual ensemble members, and the second step is the appropriate combination of outputs from the ensemble members to provide a final output.

8.4.1 Simple and Weighted Averaging Method

Linear combination of the outputs of ensemble members is one of the most popular approaches for combining different outputs. Via simple averaging or a weighted average method that considers the relative performance of each model, a single output can be created from the combination of the outputs resulted by a set of models. Combining through simple averaging is defined as

$$\hat{y}_i = 1/m\left(\sum_{j=1}^{m} \hat{y}_{ij}\right) \quad i = 1, \ldots, n \tag{8.3}$$

Despite its simplicity, the simple averaging method suffers from the problem of considering equal weighs for individual models. Obviously, the difference in the reliability of individual models is not considered in the simple averaging approach as all of them are assigned by similar weights. To overcome this shortcoming, the model of weighted averaging also known as *stacking* might be used. This method is presented by the following function:

$$\hat{y}_i = \sum_{j=1}^{m} c_j \hat{y}_{ij} \quad i = 1, \ldots, n \tag{8.4}$$

where c = the weight of each individual models. Under “stacking” an additional attempt is used to learn how to combine the models by tuning the c weights over the calibration data. To derive c weights, Shu and Burn (2004) suggested minimizing the following function over m models and n calibration data.

$$w = \sum_{i=1}^{n} \left[\frac{y_i - \sum_{j=1}^{m} c_j \hat{y}_{ij}}{y_i} \right]^2 \quad c_k > 0 \tag{8.5}$$

The stacking method uses constant weights over the time period of the calibration data set that reduces the flexibility of the method in facing different states of a system.

8.4.2 Relying on the User's Experience

In this method, at each step of decision-making, instead of combining outputs of different models, the result of just one model is selected, relying on the experience of the last step. Obviously, this method is limited to cases of time series forecasting where predictors are well correlated. Using this method is not recommended for general function approximation.

8.4.3 Using Empirical Models

Empirical models, particularly artificial neural network (ANNs), are known as powerful tools for function mapping. See and Abrahart (2001) have suggested the use of ANNs as a data fusion method. The general form of this method is

$$\hat{y}_i = g\left(\left[\hat{Y}_i\right]\right) \quad \left[\hat{Y}\right] = \begin{bmatrix} \hat{y}_{i1} \\ \hat{y}_{i2} \\ \vdots \\ \hat{y}_{im} \end{bmatrix}; \quad i = 1, \ldots, n \tag{8.6}$$

where g is a nonlinear function which maps outputs of different individual forecast models to a single output of $\hat{y}_i$ using an ANN model. Like most empirical methods, this method might suffer from the lack of statistical sense in the mechanism of data processing.

8.4.4 Using Statistical Methods

Araghinejad et al. (2011) proposed a statistical method for data fusion based on the concept of K-NN estimation algorithm which is presented as follows:

1. Use m individual forecast models to produce $n \times m$ estimation, where n is the number of observed data used for calibration.
2. Evaluate individual forecasting models in all n forecast experiences. Compute the matrix of $[A] = [a_{ij}]_{n \times m}$, where $[a_{ij}] = 1$ if mth model results in the best forecast during ith experience;
   otherwise $a_{ij} = 0$.
3. At the estimation time t, compute m forecasts of y_t using m individual forecast models and develop

$$[\hat{Y}] = \begin{bmatrix} \hat{y}_{i1} \\ \hat{y}_{i2} \\ \vdots \\ \hat{y} \end{bmatrix}$$

4. Compute $[F] = [A] \times [\hat{Y}] = \begin{bmatrix} f_1 \\ f_2 \\ \vdots \\ f_n \end{bmatrix}$
5. Use a distance function to find the distance between present time predictors, $[X_t]$, and the historical predictors, $[X_j]$. Determine the nearest neighbors from n sets of observed data, such that the smaller distance is assigned to the nearest neighbor.
6. Estimate the dependent variable by the following equation:

$$\overline{y}_r = \frac{1}{\sum_{i=1}^{k} 1/i} \sum_{i=1}^{k} (1/i) f_i \tag{8.7}$$

   where i = the order of the nearest neighbors in which the nearest have the lowest order ($i = 1$ to K), k = number of nearest neighbors, and f_i = the magnitude of nearest neighbor i.

One of the differences of this statistical model and the other ones is the use of input data not only in the individual models but also in determining the parameters of the model fusion method as shown in Fig. 8.6.

8.4.5 Individual Model Generation

Due to the flexible geometry of ANNs, they have been recognized as suitable models to be used in the ensemble techniques. ANN ensembles offer a number of advantages over a single ANN since they have the potential for improving generalization. Cannon and Whitfield (2002) and Shu and Burn (2004) used ANN ensembles in the hydrological context. Cannon and Whitfield (2002) used ANN ensembles to predict changes in streamflow conditions in British Columbia,

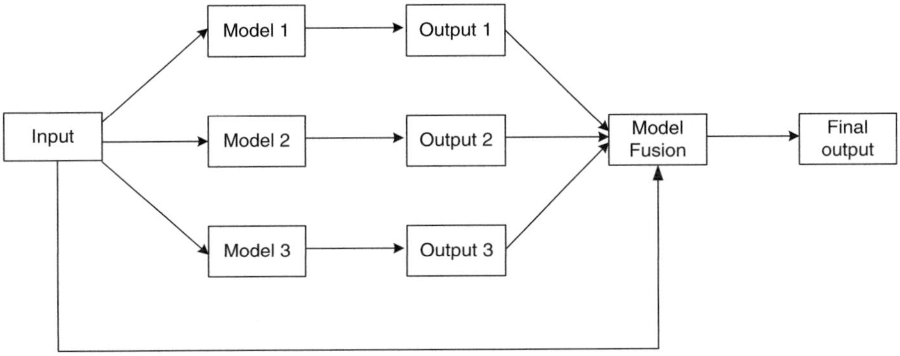

Fig. 8.6 A proposed statistical approach for multi-model data fusion

Canada, through a downscaling model. Shu and Burn (2004) applied artificial neural network ensembles in pooled flood frequency analysis for estimating the index flood and the 10-year flood quintiles. Recently, Araghinejad et al. (2011) applied ANN individual models for estimation of hydrological variables in a probabilistic manner.

Individual ANN models can be produced by the following approaches:

- Changing the objective function of an ANN
- Weighting the data in the calibration set
- Using different random numbers through the calibration of an ANN (say random initial weights and biases)
- Using selective data as calibration data set

References

Abrahart R, See L (2002) Multi-model data fusion for river flow forecasting: an evaluation of six alternative methods based on two contrasting catchments. Hydrol Earth Syst Sci 6(4):655–670

Araghinejad S, Azmi M, Kholghi M (2011) Application of artificial neural network ensembles in probabilistic hydrological forecasting. J Hydrol 407(1–4):94–104. doi:10.1016/j.jhydrol.2011.07.011

Cannon AJ, Whitfield PH (2002) Downscaling recent streamflow conditions in British Columbia, Canada, using ensemble neural network models. J Hydrol 259(1–4):136–151

See L, Abrahart RJ (2001) Multi-model data fusion for hydrological forecasting. Comput Geosci 27:987–994

Shu C, Burn DH (2004) Artificial neural network ensembles and their application in pooled flood frequency analysis. Water Resour Res 40:W09301. doi:10.1029/ 2003WR002816

Appendix

A. Basic Commands in MATLAB

A.1 Introduction

MATLAB is a numerical computing environment developed by MathWorks, for matrix manipulations, plotting of functions and data, implementation of algorithms, creation of user interfaces, and interfacing with programs written in other languages. The name of MATLAB, "MATrix LABoratory," refers to the beginning objective of developing it, which was a tool for dealing with matrix calculations.

To get started with the MATLAB, a brief review on the different sections of its main page as presented in Fig. A.1 might be useful.

Typically the main page of MATLAB is consisted of seven sections which are defined as follows:

1. Different menus are called from the menu bar as shown in section 1 of Fig. A.1
2. The icon shown in section 2 is used to create an *m* file (which is going to be discussed later).
3. Section 3 demonstrates the address, where different files are restored or loaded.
4. Section 4 is the platform where different programs are recalled, and run, and the process of calculation is reported. Small programs which are not going to be stored could be written in this section.
5. Section 5 contains necessary information on the preprepared toolboxes of MATLAB.
6. Section 6 stores and reports numerical input and output variables. It is actually a spreadsheet within MATLAB.
7. In section 7 the commands are stored to save the history of the operation. Users can call previous commands used from starting MATLAB.

MATLAB is a case-sensitive program which means it makes a difference if one uses capital or small letters. This simple but very important rule should be considered in programming by MATLAB. The variables could be one number or a matrix of

S. Araghinejad, *Data-Driven Modeling: Using MATLAB® in Water Resources and Environmental Engineering*, Water Science and Technology Library 67,
DOI 10.1007/978-94-007-7506-0, © Springer Science+Business Media Dordrecht 2014

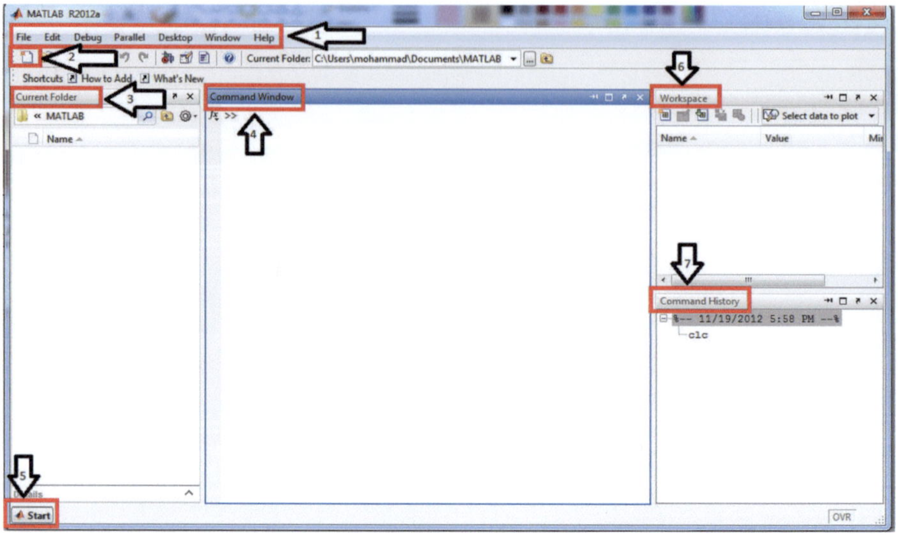

Fig. A.1 A typical main window of MATLAB

various dimensions. To create a matrix, "space" or "," is used to separate different columns, where ";" or "enter" is used to separate rows. Examples are shown below.

```
b=[3 4 5; 6 7 8]
b =
       3    4    5
       6    7    8

>> c=[6,8,8
7,8,9]
c =

       6    8    8
       7    8    9
```

After introduction, basic information on different variables, operators, and functions is presented. Next an introduction on the basic commands for programming in MATLAB is reviewed. Finally, how to draw different plots is discussed.

A.2 Variables, Operators, and Functions

A.2.1 Predefined Variables

In MATLAB some names have been reserved for specific variables. Creation of a new variable with the name of a predefined variable should be avoided. The list of those variables is shown in Table A.1.

Table A.1 Predefined variables in MATLAB

Name	Variable
Inf	∞
Eps	2.2204e-16
Pi	3.1416
NaN	Undefined

A.2.2 Predefined Matrices

Like predefined variables, some names have been reserved for predefined matrices which are presented as follows.

ones() makes a matrix with "one" elements.

```
a=ones(2,3)
a=
      1   1   1
      1   1   1
```

zeros() makes a matrix with "zero" elements.

```
>>a= zeros (1,2)
a=
      0   0
```

rand() makes a matrix with random variables as elements.

```
>>a=rand(2,2)
a =
     0.8147  0.1270
     0.9058  0.9134
```

eye() makes a unit diagonal matrix.

```
>> a=eye(2,2)
a =
      1   0
      0   1
```

diag( ) develops a diagonal matrix given a vector or obtains the diagonal elements of a specific matrix. An example is presented below.

```
>> a=[2  3 4] ;
>> b=diag(a)
b =
       2   0   0
       0   3   0
       0   0   4

>> c=[ 3 4 5; 5 6 7; 7 8 9];
>> d=diag(c)
d =
       3
       6
       9
```

To call the element of a matrix, we need to mention its index. If we need to call the entire column or the entire row, ":" is used as shown in the following examples:

```
>> a=[2 3  4; 5 6 7; 7 8 9]
a =
        2   3   4
        5   6   7
        7   8   9

>> a(2,3)
ans =
        7

>> a(2)
ans =
        5

>> a(:,2)
ans =
        3
        6
        8
```

A.2.3 Basic Operators

Basic operators in MATLAB are shown in Table A.2.

The operators presented in Table A.2 are used for matrix calculation. In case of applying them in an element by element basis, one can use a point before the associated signs (*. ‹ / . ‹ ^.).

Table A.2 Basic operators in MATLAB

Operation	Sign
Sum	+
Difference	−
Multiplication	*
Division	/
Power	^

The following example shows how the basic operators are used in MATLAB:

Example A.1: Basic Operators

Create matrix *S* as the sum of matrices *b* and *c*, matrix *Z* as the element by element multiplication of matrices *b* and *c*, and *Zp* as matrix *Z* with elements powered by 2.

$$b = \begin{bmatrix} 3 & 4 & 5 \\ 6 & 7 & 8 \end{bmatrix}$$

$$c = \begin{bmatrix} 6 & 8 & 8 \\ 7 & 8 & 9 \end{bmatrix}$$

Solution

```
>>a=2;
b=[3 4 5; 6 7 8];
c=[6 8 8; 7 8 9];
S=b+c;
```

```
Z=b.*c;
```

```
Zp=z.^a;
```

Table A.3 Conditional and logical operators in MATLAB

Operator	Character
Greater than	>
Less than	<
Equal to	==
Equal and greater than	>=
Equal and less than	<=
Unequal to	~=
And	&
Or	|
Complement	~

A.2.4 Conditional and Logical Operators

Conditional and logical operators are widely used in programming, which will be presented in the next section. A list of these operators is shown in Table A.3.

A.3 Commands and Programming

A.3.1 Basic Commands

The following commands are the most basic comments used in MATLAB:

clc clears the command window page.
clear clears the contain of the workspace.
input asks the user to input a specific variable. For instance,

```
>>a=input('please enter a number:')
```

which results in

```
please enter a number: 3
a =
     3
```

We can also use the following syntax to get an input as a string:

```
>> a=input('please enter your name:','s')
```

which results in

```
please enter your name: Roz
a =
   Roz
```

disp displays the name or value of a variable. If the variable is between ' ', its name will be displayed. Otherwise its value will be displayed.

```
>> a=34;
>> disp (a)
    34

>> disp('hello')
    hello

>> disp ('a='); disp (a);
a=
    34
```

close closes plotted figures and plots.

A.3.2 *m* File

An *m* file is a file saved by extension *m* containing programs that can be run by MATLAB. Actually it is not wise to write our programs in command window because they could not be saved and used again. Creating an *m* file enables storing and sharing our programs. The rules of programming in an *m* file are similar to those of command window. Generally, the variables in the left side of an equation can be named by the user, and those in the right hand of an equation should be either introduced in previous commands or be predefined in MATLAB. Usually we use several functions within an *m* file. A function may be predefined in MATLAB or be created by the user as will be described in the next section.

The following are some comments that might be used through the programming in an *m* file. It will be very useful to insert comments within an *m* file to describe different section of a program. Avoiding comments might cause confusion in the understanding of the content of a program even for the programmer himself. Starting a line with % turns it to a comment, which does not appear during the run of a program. It is better to save the *m* files in a specific address (preferably the current folder). The user can then open the files by the open button in the menu bar. Do not use negative sign (−) immediately after an equation. One may use parenthesis to use this sign in the programming.

The following example demonstrates a simple *m* file.

Example A.2: A Simple m File

Write a program to calculate the area and perimeter of a rectangle by its length and width.

Solution

```
%The program calculates the area and perimeter of a
rectangle
a= input('Length of rectangle');
b= input('Width of rectangle');
Area=a*b;
perimeter= 2*(a+b);
disp ('Area='); disp(Area);
disp ('perimeter='); disp (perimeter);
```

The above program can be saved as "rectangle.m."

A.3.3 Functions

Most of the functions we use in programming are actually predefined functions in MATLAB. The general form of a function is as follows:

function (outputs) = "name of the function" (inputs)

.

.

.

end

A function should be saved by the name of function. To execute a function one can call the name of the function and its input variables. Also, to execute a function, the following command can be used.

fval('name of the function', 'input 1', 'input 2', . . .)

Example A.3: Function Development

Write a function to calculate the area of a circle by its radius.

Solution
The following function is developed as an *m* file and is saved in the "current folder":

```
function [Area,Perimeter]=func1(Radius)
Area=(Radius^2)*pi;
disp('Area='); disp(Area);
Perimeter =2*Radius*pi;
disp('Perimeter='); disp(Perimeter);
end
```

Then by typing "func1(radius)", the area and perimeter of the circle with the specific radius is calculated.

Useful Predefined Functions

These are some useful predefined functions in MATLAB.

max( ) returns the maximum element of columns of a matrix.

```
>> a=[1 2  3 4 5];
>> max(a)
ans =
     5
```

min( ) returns the minimum element of columns of a matrix.

```
>> a=[1 2  3 4 5];
>> min(a)
ans =
     1
```

sum( ) returns the summation of columns of a matrix.

```
>> a=[1 2  3 4 5];
>> sum(a)
ans =
    15
```

mean( ) returns the mean of elements of columns of a matrix.

```
>> a=[1 2  3 4 5];
>> mean(a)
ans =
      3
```

All the above functions return the output for each column. To apply the function on all elements, we need to use the function twice. `sum(sum(a))` and `mean(mean(a))` are two examples of that approach.

length( ) returns the number of elements in a vector. In case of a matrix of $m \times n$, the maximum values of m and n is returned.

```
>> a=[  7 5 4 3];
>> b=length(a)
    b =
        4
>> c=[ 7 5 5 6;5 4 4 3];
>> d=length(c)
    d =
        4
```

abs( ) calculates the absolute value of the elements of columns of a matrix.

```
>> a=[2 -3  4;4 -4 2];
>> b=abs(a)
b =
      2   3   4
      4   4   2
```

size( ) returns number of rows and columns of a matrix.

```
>> a=[2 3  4;5 6 7];
>> [m n]=size(a)
m =
      2
n =
      3
```

Table A.4 Predefined mathematical functions in MATLAB

Symbols	Function
exp( )	Exponential
log( )	Logarithm
log10( )	Logarithm to base 10
sign( )	Sign function
sin( )	Sinusoidal
cos( )	Cosine
tan( )	Tangent
cot( )	Cotangent

sqrt( ) calculates the square root of the elements of columns of a matrix.

```
>> a=[2 4 9];
>> b=sqrt(a)
b =
     1.4142   2.0000   3.0000
```

find( ) finds location of specific values in a vector.

```
>> a=[2 3 4 -4 5 6 0];
>> find(a<=0)
ans =
      4   7
```

sort( ) sorts elements of a vector from lowest to highest.

```
>> a=[2 3 4 -4 5 6 0];
>> b=sort(a)
b =
     -4   0   2   3   4   5   6
```

Table A.4 shows the other famous predefined mathematical functions in MATLAB.

If "a" is used before a mathematical function (say sin), it is changed to the inverse function (asin). If "h" is used after a function, its hyperbolic will be used (sinh).

A.3.4 Conditional Commands

If, switch, for, and while are categorized as conditional commands which are described as follows:

if

The general form of an "if" command is as follows:

If condition 1
commands
else If condition 2
commands
else
end
end

An example of using "if" command is presented in Example A.4.

Example A.4: Using "if" Command

Develop a program to calculate the benefit of a company which is calculated based on the sold products as follows.

Benefit = 1,000 × number of products if sold products are less than 1,000
Benefit = 700 × number of products if sold products are between 100 and 1,000
Benefit = 1,200 × number of products if sold products are more than 1,000

Solution

```
n= input(' Number of product');
  if n>1000
    benefit=n*1200 ;
  else if 100<n<1000
      benefit=700*n;
    else
            benefit=100 *n;
    end
  end
```

Switch

Switch is used in cases where several conditions are used. General form of switch is as follows:

switch *name of the variable*
case *value of the variable*
commands
case *value of the variable*
commands
otherwise
commands
end

Example A.5: Using Switch Command

Develop a program to get *a* and multiply it by 2 if a is equal to 2; multiply it by 20 if a is equal to 3; otherwise multiply it by 30. Store the result in *b*.

```
a=3;
switch a
  case 2
      b=a*10 ;
  case 3
      b=a*20;
  otherwise
      b=a*30;
end
```

---

for

The general structure of "for" loop is presented as follows:

for i=a:b
body of loop
end

The above form is repeated for different values of *i* starting from *a* and ending to *b* with steps equal to one. In case of changing the steps to *c*, i=a:b is changed to i=a:c:b.

Example A.6: Developing Loops by "for"

Develop a multiplication table.

Solution

```
for i=1:9
  for j=1:9
      a(i,j)=j*i;
  end
end
```

---

while

while is used when a loop continues until a condition is satisfied. The general structure of "for" loop is presented as follows:

while "condition"
body of loop
end

Example A.7: Developing Loops by "while"

Write a program to add 1 to variable x until it is exceeds 10.

```
x=0;
while x<10
    x=x+1;
end
```

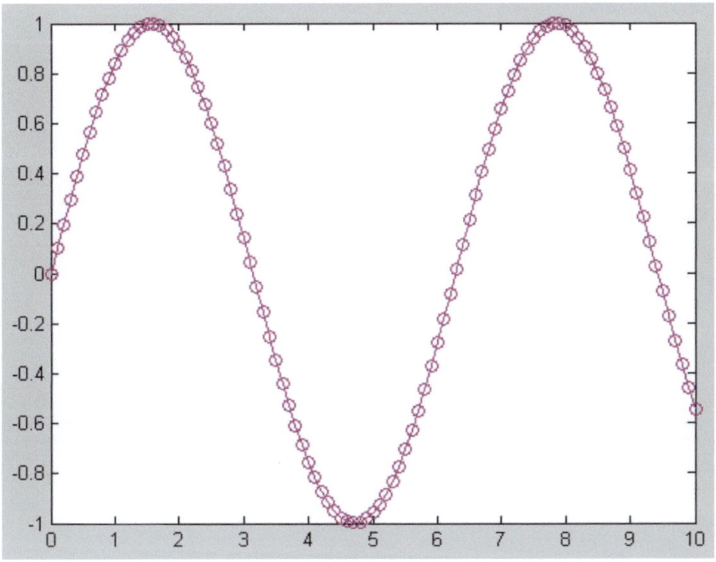

Fig. A.2 Plot of Example A.8

A.4 Plotting

A.4.1 Two-Dimensional Plot

A line or a two-dimensional curve in *xy* coordinate plane can be plotted by

```
Plot(x,y)
```

Also the color, type, and symbols of the plot can be determined by

```
plot(x,y,'om-');
```

where "*o*" determines symbol, "*m*" determines color, and "–" determines type of the plot.

Example A.8: Two-Dimensional Plot

Plot the function of $y = \sin x$ in the range of $X = [1,10]$ (Fig. A.2).

Solution

```
x=0:0.1:10;
y=sin(x);
```

```
plot(x,y,'om-');
```

A.4.2 Commenting and Labeling a Plot

Labels, titles, text on the curve, and legend for a curve are provided by the following commands:

```
xlabel (' text')
```

writes a text on x-axis.

```
ylabel (' text ')
```

writes a text on y-axis.

```
title (' title')
```

writes title of a curve.

```
text (x0,y0,'text on the curve')
```

writes text on x0,y0 coordinate.

```
legend ('comment',m)
```

provides a comment on the *m*th quarter of the page.

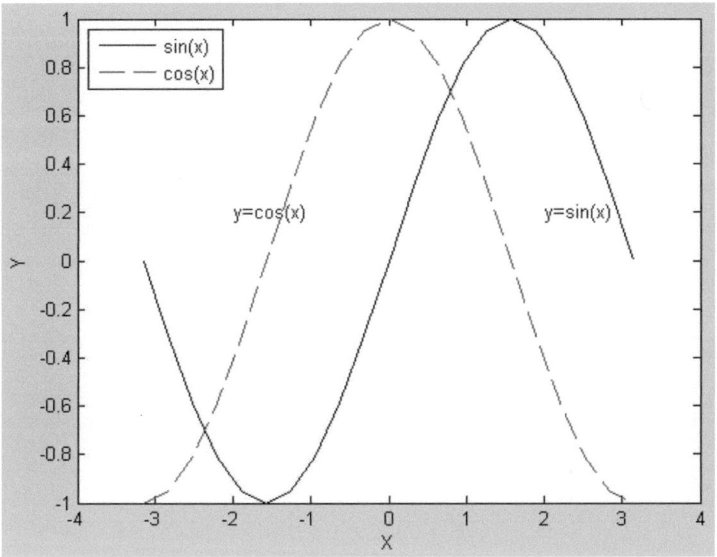

Fig. A.3 Plot of Example A.9

A.4.3 Drawing Multiple Curves on a Plot

Using the following form of plot command, one can plot multiple curves on an individual plot:

```
plot (x1,y1,'s', x2,y2,'s', x3,y3,'s',...);
```

"s" demonstrates the characteristics of each curve.

Example A.9: Multiple Curves on a Plot

Plot *sin* and *cosine* functions in the range of $[-\pi,\pi]$ (Fig. A.3).

Solution

```
x=-pi:pi/10:pi;
y1=sin(x);
y2=cos(x);
plot(x,y1,'k-',x,y2,'m-');
xlabel 'X';
ylabel 'Y';
```

(continued)

```
legend ('sin(x)','cos(x)',2);
text (2,0.2,'y=sin(x)');
text(-2,0.2,'y=cos(x)');
```

A.4.4 Drawing Multiple Curves on a Page

To draw multiple plots on a page, the following command is used:

```
subplot(a,b,c)
```

where a is number of rows, b is number of columns, and c is the index of each cell from top left.

Example A.10: Multiple Plots in a Page

Plot four curves of $y = \cos(x)$, $y = \cos(2x)$, $y = \cos(3x)$, and $y = \cos(4X)$ on a page (Fig. A.4).

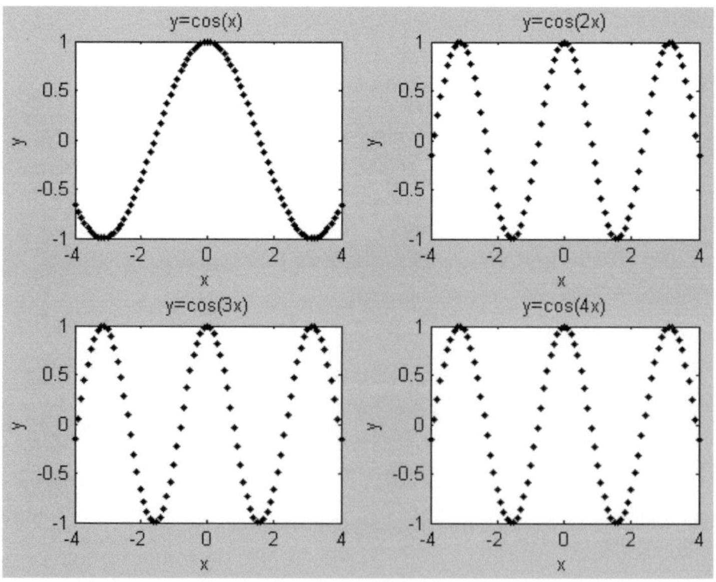

Fig. A.4 Plot of Example A.10

Solution

```
x=(-4:0.1:4);

y1=cos(x);
y2=cos(2*x);
y3=cos(3*x);
y4=cos(4*x);

subplot  (2,2,1);  plot(x,y1,'k.');title  'y=cos(x)';
xlabel 'x';ylabel 'y';
subplot  (2,2,2); plot(x,y2,'k.');title  'y=cos(2x)';
xlabel 'x';ylabel 'y';
subplot  (2,2,3); plot(x,y2,'k.');title  'y=cos(3x)';
xlabel 'x';ylabel 'y';
subplot  (2,2,4); plot(x,y2,'k.');title  'y=cos(4x)';
xlabel 'x';ylabel 'y';
```

A.4.5 Drawing Logarithmic and Semilogarithmic Curves

Sometimes we need to plot semilogarithmic and logarithmic plots. In this case `semilogy` and `semilogx` are used to plot curves which have logarithmic y-axis and logarithmic x-axis, respectively. In case of plotting logarithmic plots, `loglog` command is used.

Example A.11: Two-Dimensional Plot

For $x = 1{:}100$ and $y = \exp(x)$, plot a semilog (y-axis) and loglog plot (Fig. A.5).

Solution

```
x=1:100;
y=exp(x);
subplot(1,2,1); semilogy(x,y); xlabel 'x'; ylabel 'log
y'; title 'semilog';
subplot(1,2,2); loglog(x,y); xlabel 'log x'; ylabel
'log y'; title 'loglog';
```

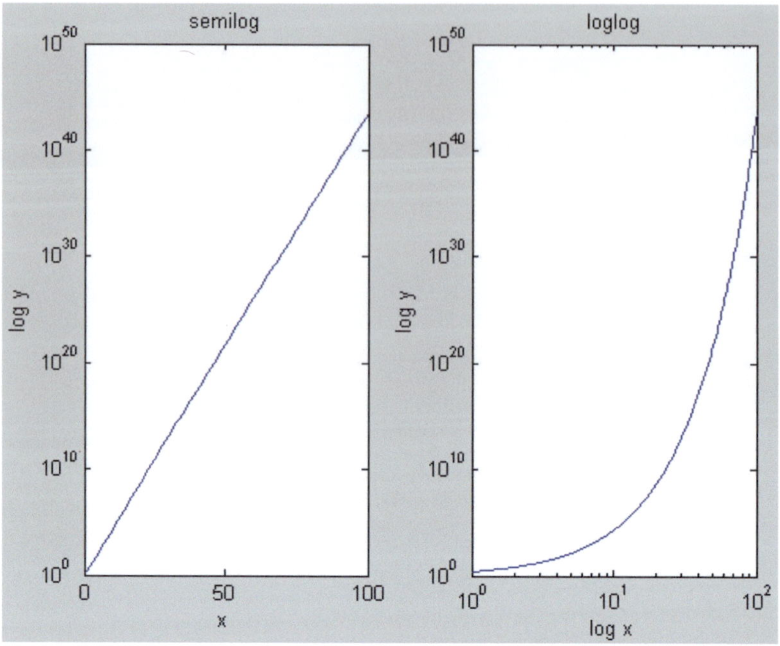

Fig. A.5 Plot of Example A.11

A.4.6 Three-Dimensional Plot

To plot a three-dimensional curve, `plot3(x,y,z)` is used as seen in Example A.12.

Example A.12: Three-Dimensional Plot

Plot the following curve in range of $t = [-40{,}40]$ (Fig. A.6).

$$f(x,y,z) = \begin{cases} x = \sin(t) \\ y = \cos(t) \\ z = \sin(t) + \cos(t) \end{cases}$$

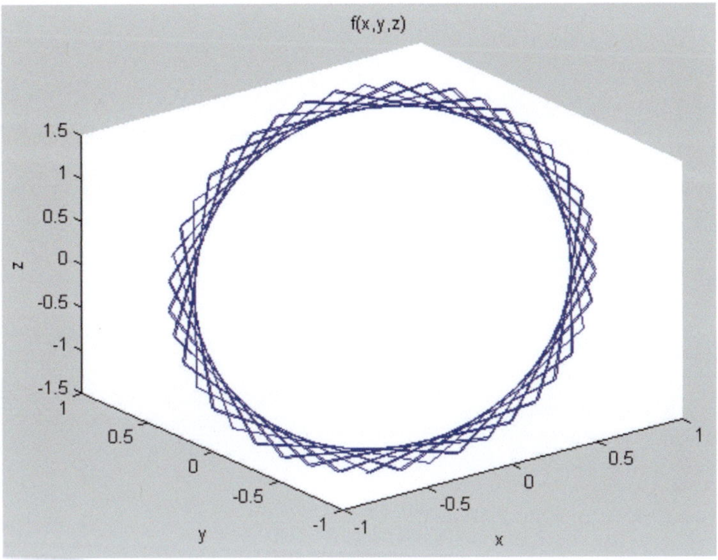

Fig. A.6 Plot of Example A.12

Solution

```
t=(-40:40);
x= cos(t);
y=sin(t);
z=sin(t)+cos(t);
plot3(x,y,z);
xlabel 'x'; ylabel 'y';zlabel 'z';title 'f(x,y,z)';
```

Index

S. Araghinejad, *Data-Driven Modeling: Using MATLAB® in Water Resources and Environmental Engineering*, Water Science and Technology Library 67,
DOI 10.1007/978-94-007-7506-0, © Springer Science+Business Media Dordrecht 2014

MIX
Papier aus verantwortungsvollen Quellen
Paper from responsible sources
FSC® C105338

If you have any concerns about our products,
you can contact us on
ProductSafety@springernature.com

In case Publisher is established outside the EU,
the EU authorized representative is:
Springer Nature Customer Service Center GmbH
Europaplatz 3, 69115 Heidelberg, Germany

Printed by Libri Plureos GmbH
in Hamburg, Germany